The New Naturalist Library

A SURVEY OF BRITISH NATURAL HISTORY

CAVES AND CAVE LIFE

The aim of this series is to interest the general reader in the wildlife of Britain by recapturing the enquiring spirit of the old naturalists. The editors believe that the natural pride of the British public in the native flora and fauna, to which must be added concern for their conservation, is best fostered by maintaining a high standard of accuracy combined with clarity of exposition in presenting the results of modern scientific research.

A cave spider *Porhomma convexum* stalks across the floor of GB Cave on Mendip. Above it, a fungus gnat *Speolepta leptogaster* hangs in a spreading fungal mycelium. (Philip Chapman)

32

The New Naturalist

CAVES AND CAVE LIFE

Philip Chapman

With 97 black and white
photographs, line drawings and maps

HarperCollins*Publishers*

HarperCollins*Publishers*
London · Glasgow · Sydney · Auckland
Toronto · Johannesburg

Dedicated to the memory of
Dr G.T. 'Jeff' Jefferson
a dear colleague and friend

First published 1993

ISBN 0 00 219908 4 (Paperback)
ISBN 0 00 219907 6 (Hardback)

Printed and bound by Butler & Tanner, Frome,
Somerset, UK

Contents

Editors' Preface

To man, caves are the original shelters, sought since Palaeolithic times hundreds of thousands of years ago. To the adventurous amongst us, they are challenges to be explored, dark passages leading to unknown underground palaces and waterways, sometimes of amazing beauty. To the naturalists amongst us, they arouse our interest by their curious and unique life-forms, selected by the restrictive environments, and by their presence in areas of limestone country of outstanding beauty. Yet, as with other life, the plant and animal communities of caves form a cohesive and interacting collection of organisms, from bacteria to mammals, from lower to higher plants, depending on the varied local environments within the cave systems.

Here, then, is an ideal subject for the *New Naturalist,* taking into account not only the living natural history of caves, but also their origin, habitat characteristics, and what they tell us of past times. Indeed, as well as their living content of caves, the sediments within them are often the graveyard of past denizens of caves, such as the hyaena, as well as the prey of cave carnivores; and, of course, these sediments reveal past habitation by man through the present of bones and tools. So we have a fourth dimension of time to add to the natural history of caves.

It may be thought that cave communities would be one of the few remaining natural ecosystems surviving in the British Isles, protected by difficulty of access. As with other living communities more apparent and better known to us, this is not the case; they are perhaps more fragile than above ground communities, more easily disturbed and affected by man's activities. To the natural historian the subject of caves demands a broad multidisciplinary approach. Dr Chapman has extensive experience of the many aspects of cave natural history. He has been able to integrate this variety, dealing with the essential geological and geomorphological background, the historical theme, and the natural history of caves, so presenting the naturalist with an outstanding and cohesive account of a unique and extraordinary ecosystem of wide interest.

Acknowledgements

Mavis Jefferson kindly allowed me to peruse and quote from the papers and notes of her late husband Dr G.T. 'Jeff' Jefferson, to whose memory I gratefully dedicate this book. Chris Howes contributed most of the photographs contained in the text and prepared all the black and white prints. Charlie Self contributed sections on the caves of Co. Clare and Mendip. Pam Fogg contributed the section on the caves of Fermanagh. Mick McHale contributed the section on the caves of Derbyshire, which was critically reviewed by John Beck. Martyn Farr critically reviewed the section on the caves of South Wales and contributed several photographs. Dave Brook critically reviewed the section on the caves of the Northern Pennines. Graham Proudlove contributed a wealth of material on cave fishes. Richard Wright of NCC and John Gunn of the Limestone Research Group at Manchester Polytechnic (now Huddersfield University) kindly allowed me to refer to and quote extensively from an NCC-commissioned report on the impact of agricultural operations on the scientific interest of cave SSSIs, as well as supplying additional information relating to the conservation of caves. John Gunn then critically reviewed the resultant Chapter 7 which is based largely on his and Paul Hardwick's work. David Judson supplied valuable information relating to cave conservation. Trevor Shaw critically reviewed Chapter 1 and provided most of the illustrations from his own collection. Tony Waltham supplied material on cave and karst SSSIs, and critically reviewed the geological information contained in Chapter 2. Christopher Stringer of the Natural History Museum, London, critically reviewed Chapter 6. Peter Glanvill and Mark Woombs critically reviewed the section on submarine caves which was compiled from the unpublished results of their researches and those of Rob Palmer, Bernard Picton, Richard Park, Richard White and John White. Cliff Davies contributed the results of his research on Swallows nesting in caves and reviewed the resulting section. Rowley Snazell of ITE's Furzebrook Research Station and Paul Hillyard of the Natural History Museum, London, contributed material on cave and chalkland spiders. K L Brown told the Olm story to Mark Day who passed it on to me. Michael Woods of Cheddar gave information on badgers in caves. Tony Boycott contributed material on Green Holes, badgers in caves and Leptospirosis. Paul Harding supplied distributional information held by ITE at Monks Wood Experimental Station. Bob Stebbings supplied information on bats in caves. To all those who helped to make this book possible, and particularly Brin Edwards whose superb illustrations appear throughout the text, and to my editors at Collins, Elizabeth Stubbs and Myles Archibald, I offer my sincere thanks.

Phil Chapman, 1991.

1

The Fascination of Caves

The lure of caves

There is a curious fascination about caves that seems to affect people of all ages and all cultures. Even as children, we have a kind of longing for caves, seeing them perhaps as a place of safety, but equally as a source of adventure and excitement – a gateway to the unknown.

Our remote ancestors used the entrances of caves as habitations, but reserved their depths as hiding places for their most precious and powerful secrets – the painted, magical symbols which would ensure a continuing supply of game for hunting, and the earthly remains of their dead. Religion was born in caves, and even now the buildings of our Christian cultures retain atavisms of those earlier forms of worship; under the central part of the church lies the crypt, secret and dark – originally the burial place of saints and martyrs. It is perhaps also significant that the Mother of God should have appeared to Bernadette in a grotto at Lourdes, and should have consecrated the cave spring which welled up from underground.

In Japanese mythology the sun-goddess Amaterasu retreated at night to a cave, plunging the world into darkness. The ancient Greeks too gave prominence to caves in their mythology. Zeus, chief of Gods, was born in a cave, and of course the Greek hell lay below ground, and at its gates Charon the ferryman waited in his boat to row the souls of the departed across the black waters of the River Styx into a land of grief and eternal pain. In our own mythology, King Arthur, his knights and hounds are said to slumber still beneath a Welsh mountain, eternally awaiting the call to battle. Even today in parts of New Guinea, tribesmen will say that their ancestors were born directly from the earth through the womb-like opening of a cave.

With such a long cultural association between the darkness of caves, their chill and smell of decay, and the nameless terrors of the grave, it is not surprising that our ancestors should have equated caves with what they knew of volcanic vents and imagined the fires of hell in their depths. Dante's *Inferno* is just one manifestation of an older oral tradition in Europe which told of animals, usually a dog or a goose, entering a cave to emerge days later from another miles away, devoid of fur or feathers and showing signs of singeing by infernal flames.

Lurid accounts of real caves are frequent in ancient literature. The Roman philosopher Seneca reported that a party of Greek silver prospectors who ventured underground had encountered:

> "huge rushing rivers, vast still lakes, and spectacles fit to make them shake with horror. The land hung above their heads and the winds whistled hollowly in the shadows. In the depths, the frightful rivers led nowhere into the perpetual and alien night."

Seneca adds that after their return to the surface, the miners "lived in fear for having tempted the fires of Hell."

Fig. 1.1 Manner of crossing the first river in Peak Cavern, engraved by Cruikshank in 1797, from G.M. Woodward's *Eccentric excursions ... in different parts of England & South Wales*, pub. Allen & Co., London, 1801. (Courtesy of Trevor Shaw)

Perhaps the oldest written reference to a cave appears in a book about mountains written before 221 B.C. in China – a country where caves have been systematically explored and exploited over many centuries as water supplies and as sources of nitrates for fertilizer and for making gunpowder. The earliest surviving reference to a cave in Britain dates from around 200 A.D. in the writings of Titus Flavius Clemens, known as Clement of Alexandria. He writes:

"Such as have composed histories concerning the Britannic islands tell of a cavern beneath a mountain, and at the summit of it a cleft, and of how from the wind rushing into this cavern and reverberating from its hollows, an echo as of many cymbals is heard."

Which cave this refers to is uncertain, but a location in either the Mendip Hills or Derbyshire would seem most likely since both areas were well-known as important centres of lead mining during this period. Current opinion favours the Great Cave of Wookey Hole which would undoubtedly have been known in Roman Britain and where, according to Balch (1929), a noise like the clash of cymbals can occasionally be heard.

Irish caves were also documented from the earliest times. The *Annals of the Four Masters*, written in 928 A.D., record the massacre of 1000 people in Dunmore Cave in County Kilkenny and if the abundant remains excavated there are anything to go by, the account may well be true.

The myth of the 'howling cave' resurfaces with Henry of Huntingdon in his mediaeval *Historia Anglorum*, written in Latin around 1135. He gives pride of place among the four "wonders of England" to a cave "from which the winds issue with great violence". This one appears to have been situated in the Peak District and may have been Peak Cavern, since some time later Gervase of Tilbury, writing about this cave around 1211, states that strong winds sometimes blow out of it. The third of Huntingdon's four wonders was also a cave, this time one situated at:

"Chederole where there is a cavity under the earth which many have often entered and where, although they have traversed great expanses of earth and rivers, they could never come to the end."

This poses modern scholars with an interesting puzzle, for although there is indeed a well-known cave at Cheddar (open to the public as 'Gough's Cave'), it is short, easily explored and does not connect with the underground river known to flow beneath it. However, in 1985, cave divers Rob Harper and Richard Stevenson squeezed down a narrow pit in a forgotten corner of Gough's Cave and emerged underwater into the main river, which they followed upstream to reach a large dry cavern, dubbed the Bishop's Palace. The flooded system lies close to the water table, and the divers surmise that before the Cheddar rising was enclosed by a dam, the water level may have been low enough to permit entry into the now-flooded cave. On the other hand, the village of Cheddar (once known as 'Cheddrehola') is only some ten kilometres away from Wookey Hole, and Huntingdon may well have confused the two localities.

The surprising thing is not the doubts about the accuracy of Henry of Huntingdon's accounts, but that he should have chosen caves for two of his four 'wonders of England'. Other mediaeval chroniclers also mention caves and mostly follow Huntingdon's accounts closely, as also do the various manuscript 'Wonders of Britain' or 'Mirabilia' which appeared between the 13th and 15th centuries.

In the 16th and early 17th centuries writers such as Leland, Camden, Drayton and Leigh, gripped by the Elizabethan romantic passion for 'discovering the countryside', penned lurid accounts of the caves they visited and of the legends and folklore attached to them. One looked forward to a planned visit to Wookey Hole with not a little trepidation:

"though we entered in frolicksome and merry, yet we might return out of it Sad and Pensive, and never more be seen to Laugh whilst we lived in the world."

The early 17th century saw the rise of a craze for so-called 'rogue books' – sensationalized accounts of swashbuckling anti-heroes such as highwaymen and pirates, and some of these make reference to dastardly goings-on in the caves of Derbyshire. Sam Ridd's *The Art of Juggling* (1612) portrayed Peak Cavern as a notorious centre of knavery, and Ben Jonson makes several allusions to this cave (under a different name) and its association with beggars and vagabonds in *The Devil is an Ass* (1616) and *The Gipsies Metamorphosed* (1621).

Later in the 17th century the Peak District and its caves continued to attract attention through the writings of Charles Cotton, best known for his collaboration with Izaak Walton on later editions of *The Compleat Angler*. Cotton's fondness for caves may be not altogether unconnected with his habit of using them as a sanctuary when hiding from his creditors.

Fig. 1.2 An imaginary 'straightened out' view of Peak cavern. In the foreground are the rope-makers' cottages. From a copper engraving titled *The Devil's Arse, near Castleton, in Derbyshire* which appeared in Charles Leigh *The natural history of Lancashire, Cheshire and the Peak in Derbyshire*, pub. Oxford, 1700. (Courtesy of Trevor Shaw)

Daniel Defoe, the great traveller and polemicist, seems to have completely failed to appreciate the 'Wonders of the Peak' which so enthused his contemporaries. Dubbing them the 'wonderless wonders', he selects Peak Cavern for a particularly scornful treatment:

"... where we come to the so famed Wonder call'd, saving our good Manners, *The Devil's A--e in the Peak*; Now not withstanding the grossness of the Name given it, and that there is nothing of similitude or coherence either in Form and Figure, or any other thing between the thing signified and the thing signifying; yet we must search narrowly for any thing in it to make a *Wonder*, or even any thing so strange, or odd, or vulgar, as the Name would seem to import."

This seems a bit harsh, as the entrance to Peak Cavern is, I would have thought, impressive by any standards. On the other hand, Defoe goes right over the top in his reaction to nearby Eldon Hole: "this pothole is about a mile deep ... and ... goes directly down perpendicular into the Earth, and perhaps to the Center". It is actually 75 m deep.

Although Defoe's *Tour* was not intended to be a guide book, a series of revisions by various editors up to 1778 made it ever more like one; even going to the lengths of adding in descriptions of caves not included in the original version. The success of the *Tour* and the rise of the 'picturesque' movement in art and architecture (epitomized by the romantic landscape designs of Humphrey Repton) no doubt encouraged the early 19th century vogue of 'curious travellers' who sought out and explored previously neglected corners of the countryside in order to write about their experience. Where previously the ideal landscape had been one which showed the civilizing hand of man in formal gardens and straight avenues of trees, 'wild nature' now became fashionable. Any accessible landscape featuring dramatic cliffs and wooded gorges, crumbling ruins and, of course, caves, became a tourist attraction. A swelling tide of visitors headed for the fashionable delights of the Peak and Wookey Hole, the scars and potholes of the Yorkshire Dales, the seacaves of Scotland and Kent's Cavern at Torquay. Even the great Dr Samuel Johnson seems to have been caught up with enthusiasm for a sea cave he visited on Skye during his tour of the Hebrides in 1773.

The descriptions by the 'curious travellers' generally aimed to convey emotions of awe, wonder or terror at the beauty and power of 'wild nature'. Caves

Fig. 1.3 Fingal's Cave from a hand-coloured wood engraving by Whimper, in Anon: *Natural Phenomena*, pub. London, SPCK, 1849. (Courtesy of Trevor Shaw)

lent themselves particularly well to the Gothic imaginations of young romantic writers like Benjamin Malkin, who described a visit to the entrance of Porth-yr-Ogof in his *The Scenery, Antiquities, and Biography of South Wales* (1807):

"We penetrated about an hundred yards, as far as any glimmering of daylight from the mouth directed us: and this specimen of Stygian horror was amply sufficient to satisfy all rational curiosity. The passage over uneven rocks, with scarcely a guiding light, and in many places with a bottomless gulph directly under on the left, in a misty atmosphere from the vapour of the place and the exhaustion of a laborious walk, was not to be pleasurably continued for any length of time or distance. ... Any person who will enter this cavern ... may form a just idea ... of the classical Avernus and poetical descent into the infernal regions."

The north-country clergyman John Hutton stands out among the 'curious travellers' as someone who developed a genuine interest in caves. His *A Tour to the Caves in the environs of Ingleborough and Settle* (in various editions from 1780 onwards) contained descriptions of some two dozen caves and potholes and was the first book in Britain, and one of the first in the world, whose main purpose was to describe the natural history of caves. In spite of his liberal use of Gothic adjectives such as "horrid", "dreadful" and "terrible", a real enthusiasm for his subject comes through, and in the two later editions of his book he added a section entitled "conclusions of a philosophic nature", in which he discusses limestone geology, cavern formation and hydrology. Some of his views, particularly those on the springs and underground streams of the area, were farsighted, although others seem laughably quaint in the light of modern science. It is interesting for the modern reader to note the touchstone against which he measured his own ideas:

"I think I may say without presumption, that my theory is conformable to events as related by Moses; and my reasoning agreeable to the philosophical principles of Sir *Isaac Newton*, where they could be introduced."

The early 19th century boom in natural science, when it spread to caves, focussed initially on two fields of research which Hutton had completely overlooked – namely palaeontology and archaeology. Deposits of bones had been known from caves in mainland Europe at least as far back as the 16th century, when speculation about their nature had inclined, as might be expected, to the fantastic. Some were considered to be dragon bones, while others – the *sub-fossil* tusks of elephants or mammoths, known as 'unicorn horn' – were greatly prized for their reputed medicinal properties. Quite an industry sprang up around such deposits, and their discoverers or the owners of the caves could become rich on the proceeds.

The Victorian naturalists were the first to appreciate the antiquity of cave bone deposits and their value as a geological record of Britain's past. The best known of the early cave excavators was Dean Buckland, who plundered cave deposits throughout the country in the 1820s. In the interpretation of his results he was limited by the thinking of his day, but he recognized that many of the bones were from animals no longer present in Britain, and in some cases from animals that no longer existed at all. He was the first to suggest that the explanation for the accumulation of fossil bones in some caves might be that in the distant past hyaenas had used the caves as dens and dragged the carcasses of animals into them. Buckland also made the first find of an *Old Stone*

Fig. 1.4 Gough's Cave digging in early 1935, first published in *Bristol Evening World*, 1 Feb 1935. (Trevor Shaw collection, courtesy of Cheddar Caves Ltd.)

Age human burial, the 'Red Lady of Paviland' found in Goat Hole on the Gower coast, so-called because her body had been anointed with ochre before burial.

In the 1830s, Schmerling recognized that the remains of humans and of extinct mammals found together in the same deposits in Belgium were of the same age. It was not, however, until later in the century, that the work of William Pengelly and Boyd Dawkins and their colleagues established that these remains dated back thousands of years to the Ice Ages of the Pleistocene era, when cave-dwelling people shared our familiar countryside with a fearsome array of giant animals, including cave bears, hyaenas, woolly mammoth, bison, aurochs and woolly rhinoceros. Excavation in caves has, of course, continued to the present day, and cave sites worldwide have now yielded material which has helped to shape our understanding of human evolution and the birth of our culture.

The sporting science

The systematic exploration, documentation and commercial exploitation of caves was already underway in China many centuries before miners and natural historians first began to measure and record details of our European cave

Fig. 1.5 Interior Chamber of Cox's Stalactite Cavern, Cheddar, Somersetshire. Lithograph by Newman & Co. London; pub. S. Cox, Cheddar about 1850. (Courtesy of Trevor Shaw)

systems in a scientific way. One of the first of this breed of explorers in Britain was John Beaumont, a 17th century Somerset surgeon and an amateur student of mining and geology. When in 1674 lead miners excavating a shaft in the Mendip Hills accidentally breached a natural underground chamber, Beaumont hastened to the site and hired six miners to accompany him into the cave. Carrying candles, the company descended the 18 m shaft to the first chamber, which Beaumont proceeded to measure: it was 73 m long, two metres wide and nine metres high. "The floor of it is full of loose rocks," wrote

Beaumont in his subsequent report to the Royal Society, "its roof is firmly vaulted with limestone rocks, having flowers of all colours hanging from them which present a most beautiful object to the eye." The intrepid surgeon then led a 100 m crawl through a further low passage which opened into the side of a second chamber, so vast, Beaumont reported, that "by the light of our candles we could not fully discern the roof, floor, nor sides of it." The miners, accustomed as they were to the underground, could not be persuaded to enter this chasm, even for double pay. So Beaumont went down himself:

> "I fastened a cord about me, and ordered them to let me down gently. But being down about two fathom I found the rocks to bear away, so that I could touch nothing to guide myself by, and the rope began to turn round very fast, whereupon I ordered the miners to let me down as quick as they could."

He landed dizzy but safe 21 m below, on the floor of a cavern 35 m in diameter and nearly 37 m high within which he found large veins of lead ore. Surprisingly, Beaumont's account failed to stimulate much curiosity about caves in scientific circles, and after a brief flurry of lead mining, the cave, known as Lamb Leer, was abandoned; its entrance shaft eventually collapsed and the sealed-off chamber was virtually forgotten for two centuries.

The rediscovery of Beaumont's long-lost Lamb Leer cave took place in 1879, the same year that a young French law student, Edouard-Alfred Martel, made his first visit to the famous Adelsberg Cave in Slovenia (which was then part of Austria, but since World War II has reverted to its local name of Postojna Jama). Martel was completely enthralled, and in 1883 began to devote all his vacation time to cave explorations in the Causses of southwestern France. What set him apart from previous cave explorers, was his meticulous preparations and his systematic recording of all aspects of the caves he explored, combined with a tremendous physical ability and courage. His speciality was deep vertical pits, and in 1889 he successfully negotiated the 213 m vertical entrance shaft of the Rabanel pothole north-west of Marseilles – an outstanding feat given the equipment then available.

To calculate a pit's depth, he would read the barometric pressure at the bottom and compare it with the surface pressure. He measured the horizontal dimensions of each newly-discovered chamber with a metal tape, drawing a sketch of the cave as he worked. Roof heights were calculated with an ingenious contraption: after attaching a silk thread to a small paper balloon, he would suspend an alcohol-soaked sponge beneath it, light the sponge, and measure off the length of thread carried aloft by the miniature hot-air balloon. Martel also habitually recorded subterranean air and water temperatures, finding variations with depth and season, and amassed whole volumes about cave geology, hydrology, meteorology and flora and fauna. But perhaps his greatest contribution to cave science was his research on how subterranean water circulates – a study prompted by his own bout with ptomaine poisoning, contracted from drinking spring water in 1891. After recovering from the illness, he traced the spring's source using fluorescein dye introduced to nearby sink holes. Descending the appropriate pit, he found the putrefying carcass of a dead calf that had contaminated the spring with what he wryly termed "veal bouillon". Further study allowed him to distinguish between "true springs", fed by diffuse circulation of rainwater, cleaned and filtered by its passage through soil and rocks, and "false springs" fed by a rapid flow from sinkholes via cave passages too large to filter out impurities. Martel's subsequent campaigning for stricter control of sources of drinking water eventually led to a dramatic reduction in deaths from typhoid and won him a gold medal from the French Government.

In 1895 Martel founded the French 'Société de Spéléologie', arguing that '*speleology*', which had been previously considered a sport or a singular eccentricity, should be recognized as a fully-fledged science – "a subdivision of physical geography, like limnology for lakes and oceanography for seas." A prolific author, he edited *Spelunca*, his Society's bulletin, and wrote books about his own cave discoveries. In 1907 the French Academy of Sciences awarded him the grand prize for physical sciences, and in 1928 he was elected president of the Geographical Society of Paris. By the time he died in 1938, aged 78, the grand old man of speleology had personally probed nearly 1500 caves, hundreds of which had never been entered before; his technical innovations had become standard equipment for other cavers; and above all, his persistence and dedication had created a framework within which the seedling science of speleology could develop and blossom.

Meanwhile, the systematic documentation of caves had also started in Britain, with the formation of the Yorkshire Ramblers' Club in 1892. Under the influence of S.W. Cuttriss, dubbed 'the scientist' for his assiduity in recording the group's findings, its members drew up surveys of the caves they explored, and kept notes of temperatures, altitudes and geographical features.

The great exploration challenge of the day was the awesome Gaping Gill, a pothole high on the slopes of Ingleborough Hill which had been plumbed to a depth of 110 m, but had never been descended. The main obstacle to its exploration came from Fell Beck, an icy stream which cascades down the entrance, filling it with spray and extinguishing any flame which might light a caver's descent, as well as half-drowning him. A local man, John Birkbeck, had made two heroic, but unsuccessful attempts to descend the pit in the 1840s, after digging a trench to divert the Fell Beck to another sink. His first try nearly proved fatal, when strands of his rope were severed on a rock ledge, but on his second attempt he reached a ledge at 58 m which now bears his name. Yorkshire Ramblers' member Edward Calvert took up the challenge and had almost completed his own preparations for a descent on rope ladders in 1895, when Martel arrived on the scene, hot-foot from London where he had been invited to address an International Geographical Congress on cave-hunting methods.

As something of a celebrity, Martel was encouraged by the lord of Ingleborough Manor to have a crack at the great pothole and given the support needed to refurbish and extend Birkbeck's trench. On August 1, before an eager crowd, Martel knotted together his lengths of ladder and lowered them into the darkness. As he climbed down the four metre-wide shaft he was rapidly enveloped by "half-suffocating whirls of air and water" which soaked his clothes and the field telephone which he relied upon to communicate instructions to the back-up team who controlled his safety line from the surface. The cascade redoubled 40 m down and he had to descend through a "frigid torrent gushing from a large fissure". Pausing on Birkbeck's ledge to untangle the huge heap of rope which had lodged there, he continued into the unknown depths below. Sixteen metres on, the walls of the shaft suddenly receded and Martel found himself swinging like a pendulum near the roof of

Fig. 1.6 Gaping Gill – the main chamber showing the waterfall falling 110 metres from the surface. A photograph taken in the 1930s by Eli Simpson. (Trevor Shaw collection)

an immense chamber nearly 160 m long by 30 m high. He alighted on the floor only 23 minutes after he began the descent, and characteristically at once set about measuring and sketching his discovery. For an hour and a quarter, the Frenchman revelled in the spectacle of the "Hall of the Winds", Britain's largest underground chamber from whose roof the waters of the Fell Beck tumbled "in a great nimbus of vapour and light". There was a certain sense of nationalist triumph too: "The most pleasant feature was the thought that I had succeeded where the English had failed, and on their own ground."

Martel's descent of Gaping Gill received wide publicity and awakened an interest in the possibilities of cave exploration in other parts of Britain. A group of Derbyshire rock climbers calling themselves the Kyndwr Club started to explore the caves of that county and further afield. One of their leading spirits was Dr E.A. Baker, a native of Somerset, but at that time resident in the Midlands. He was a colourful and influential character, an academic who later became director of the School of Librarianship in the University of London, but whose interest in caving was primarily sporting.

At about this time, H.E. Balch, a young postal worker at Wells in Somerset, came across a fragment of reindeer antler in the Hyena Den near Wookey Hole and, inspired by the work of Professor (later Sir William) Boyd Dawkins, at once threw himself into a study of all kinds of archaeological and fossil cave sites on Mendip. Soon Balch had founded the Wells Natural History and Archaeological Society and started the collections which eventually grew into Wells Museum. The subsequent arrival on the Mendip scene in 1902 of Baker and his colleagues from Derbyshire led to a long and fruitful collaboration, during which many of Mendip's greatest caves were dug open and explored. Both Balch and Baker were prolific writers and their publications, spread over several decades, played a large part in stimulating an interest in caving during the early part of this century. A number of clubs began to appear which cheerfully combined a scientific and sporting approach to caving, setting a pattern which has continued to the present day. Scientifically motivated 'speleologists' still recognize their dependence on sporting cavers for much of the initial exploration and often for support when working in the more exacting situations. Many are in any case themselves sporting cavers, or were in their younger days. On the other side very few of those whose motives are primarily sporting are completely uninterested in the whys and wherefores of the natural features which provide them with their sport. They also appreciate that scientific understanding increases the chances of finding more caves.

There are few completely unexplored places anywhere on the surface of the earth and none in a country such as Britain; but caves hold out the promise, or at least the hope, of completely new discovery. Most cavers must sometimes dream of one day discovering a new cave, or an extension to a known one, and of being the first to gaze upon whatever wonders it may hold. If these are the things which provide motivation for caving as a sport, they are nearly always reinforced by at least some measure of scientific curiosity, and most cavers combine the two in varying proportions.

In the years after World War II, the popularity of caving, as of so many other active pursuits, increased enormously in many parts of the world. In Britain the 1950s and 60s in particular saw a great proliferation of caving clubs and, although some of these were ephemeral, the overall level of interest and activity has remained high. Perhaps the most important development in British caving in the early post-war years was the opening up of South Wales as a major caving region. Up until 1936, cavers had taken surprisingly little interest in the area considering that Porth-yr-Ogof had been known for hundreds of years and Dan-yr-Ogof had been discovered and explored as far as the waterfalls in 1912. It was to these two caves that experienced cavers from Yorkshire and Mendip first turned their attention in 1936 and interest in the area developed rapidly. By the time that caving came to a virtual halt with the outbreak of war in 1939, the potential of South Wales was apparent to all. In 1946 the South Wales Caving Club was formed and within a few months two of its members, Peter Harvey and Ian Nixon, dug their way into the lower end of the great Ogof Ffynnon Ddu cave system on the east side of the Upper Tawe Valley. Initially rapid exploration was halted by a series of sumps (flooded sections of passage) and it was not until 1966 that a dig in the upper levels of the cave gave access into the vast maze of OFD II. A rush of new discoveries followed in quick succession over the next year, extending the vertical range of the cave to 300 m and its total length to around 40 km, making it Britain's deepest and longest cave system. (Although cave divers have since linked up various parts of the Ease Gill system in Yorkshire to give it the number one position with an overall passage length of over 70 km).

As we have seen, cave science, and in particular the archaeological and palaeontological investigation of caves, started well before the development of caving as a sport. As early as the mid-nineteenth century the living faunas of caves were receiving fairly extensive study in mainland Europe and America by the likes of J.C. Schiodte and A.S. Packard; and following the influence of Martel, the physical aspects of speleology – geology, geomorphology and hydrology – were well established there by the turn of the century.

Underground naturalists

Evidence that early man was conscious of the existence of a subterranean fauna dates back to a remarkable engraving of a cave cricket on a bison bone discovered by Count Begouen while excavating in the Grotte des Trois Frères in the French Pyrenees. The carving is believed to be 18,000 years old, yet is sufficiently clear and detailed for the subject to be recognizeable as a *Troglophilus* species, which today is distributed from Italy to Asia Minor, but no longer inhabits France.

For the next surviving reference to cave life in Europe, we must move on to the 16th century and the observant Count Trissino, who, in a letter dated 5th March 1537, recorded what must have been a form of the cave-limited amphipod, *Niphargus*. He noted that at the far end of the Covolo di Costozza in

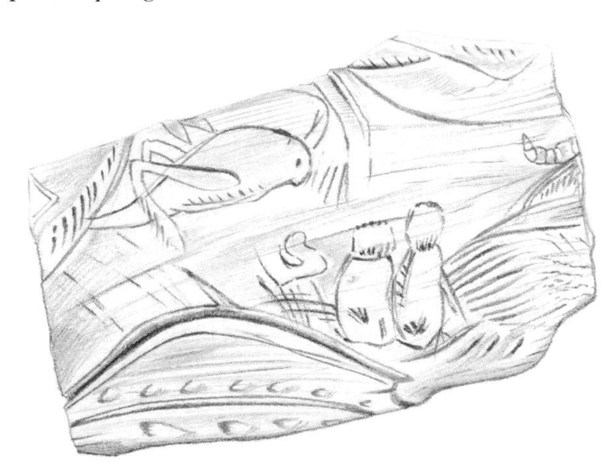

Fig. 1.7 A prehistoric engraving on a bison bone, discovered by Count Bégouen in the Grotte des Trois Frères (French Pyrenees), featuring the cave cricket *Troglophilus*. (After Bégouen)

northern Italy there was a deep pool of clear water. "In this water no fish of any kind are found, except for some tiny shrimp-like creatures similar to the marine shrimps that are sold in Venice."

The Slovenian Olm, *Proteus*, seems to have been well-known to the villagers of the Trieste area for centuries. Specimens occasionally appeared after floods in the Lintverm (from the German 'Lindvurm', meaning ' dragon') – a tributary of the River Bela near Vrihnika. With their long, pink, rather reptilian bodies, they were taken, not unreasonably, to be dragon fry – the offspring of a shadowy monster who lived in the roaring cave from which the river flowed and who caused periodic floods by opening sluice gates when her living quarters were threatened by rising water. But in the 1680s, Baron Johann Valvasor, a Slovene nobleman and well-travelled amateur scientist, ruined centuries of colourful legend by exposing the Olm as a perfectly natural blind cave salamander.

In 1799, the German naturalist-explorer Baron Alexander von Humboldt, accompanied by a French botanist called Bonpland, visited the famous Cueva del Guacharo in the Caripe Valley of Venezuela. There he collected and described a cavernicolous bird, *Steatornis caripensis*, belonging to the order which includes the nightjars, which had been known for a long time to the Indians under the name 'guacharo'. Humboldt was greatly impressed by the screeches produced by the birds when disturbed at their cave roost. "Their shrill and piercing cries strike upon the rocky vaults," he wrote, "and are repeated by the subterranean echoes." Having heard them myself, I would describe the racket as the sound of a thousand mad chickens locked up in a barn with a fox.

In 1808, Schreibers discovered the first invertebrate cave fauna in Austria and more extensive collecting was done in the Postojna area by Count Franz

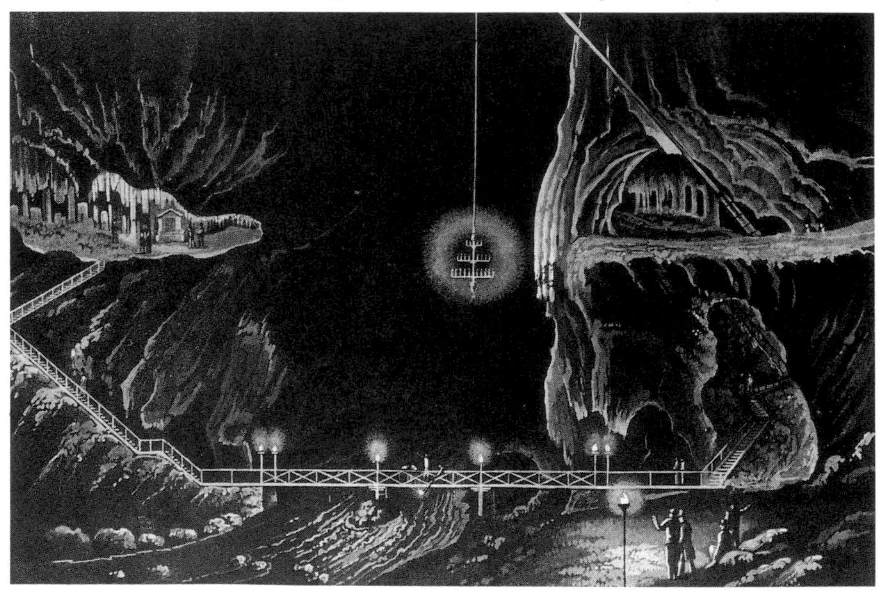

Fig. 1.8 The old route across the underground river Pivka in the Great Hall of Postojna Jama in Slovenia from an aquatint engraved by G. Dobler after a painting by Alois Schaffenrath, published in 1830. (Courtesy of Trevor Shaw)

von Hohenwart and others from 1831 onwards. It was there too that the Danish zoologist J.C. Schiodte recognized that cave faunas showed differing degrees of specialization to life in darkness, and so laid the foundations for a system of ecological classification of cave life. This was advanced in a more rigorous form in 1854 by J.R. Schiner and has been widely used by cave biologists ever since. This work perhaps marked the beginnings of the systematic discipline of '*biospéologie*', a term proposed by Armand Viré in 1904, to refer to the study of subterranean life.

In the USA important work continued intermittently from 1840. In that year Davidson collected the first specimens of a blind white fish in Mammoth Cave, described by de Kay, Wyman and Tellkampf as *Amblyopsis spelaea*. Tellkampf went on to describe other fauna from Mammoth Cave and was followed by E.D. Cope and A.S. Packard, whose remarkable studies through the 1870s put America for a time at the forefront of biospeleological research. Meanwhile, in the 1840s, V. Motschoulsky reported the first cave-specialized insects captured in the caves of Caucasia, and in 1857 De la Rouzee discovered the first cavernicolous insects known from France.

Scientific investigation of our cave faunas began in a round-about way in about 1852, when Professor Westwood and S. Bate included the following reference in their *History of the British Sessile-eyed Crustacea, Vol.1*, published in 1863.

> "In the year 1852", writes Bate, "Professor Westwood was so fortunate as to obtain from a pump-well near Maidenhead, a quantity of [*Niphargus* sp.] ... since when they have been found in Hampshire, Wiltshire ... and very recently in Dublin."

Shortly afterwards, news of the discovery in Europe and America of strange blind cave animals prompted Naturalists E. Percival Wright and A.H. Haliday to search for similar creatures in Mitchelstown New Cave in Co. Tipperary. Their search was successful and they described their find – a tiny Collembolan doubtfully identified as *Lipura stillicidii* Schiodte – in a paper read before a British Association Meeting in Dublin in 1857.

More than thirty years were to elapse before the next glimmer of enthusiasm for Irish cave life manifested itself in the form of a joint excursion in 1894 by the Dublin, Cork and Limerick Field Clubs to the Cave of Mitchelstown. One of the participants, George H. Carpenter, recorded that "after an informal luncheon on the roadside, the party being provided with candles, descended the sloping passage and ladder which led to the depths below." They spent two hours searching for cave animals and, although they failed to reach the underground river, made a reasonable collection of fauna, including the rare blind cave spider now known as *Porrhomma rosenhaueri*. In the same year, pioneering English arachnologist F.O.P. Cambridge collected spiders in Wookey Hole, but without finding anything of particular interest.

Early in 1895 E.A. Martel and his wife paid a well-publicised visit to Ireland. The event prompted the Fauna and Flora Committee of the Royal Irish Academy to support H.L. Jameson with a grant "to further investigate cave fauna in Ireland". He joined the Martels in the Enniskillen area of Co. Fermanagh and, while the Frenchman surveyed the caves and drew up his plans, Jameson collected cave animals. The interest seems to have persisted, for Jameson is also known to have made faunal collections in Speedwell Mine in Derbyshire in 1901, but there then followed a gap of over thirty years during which British cave fauna was again neglected.

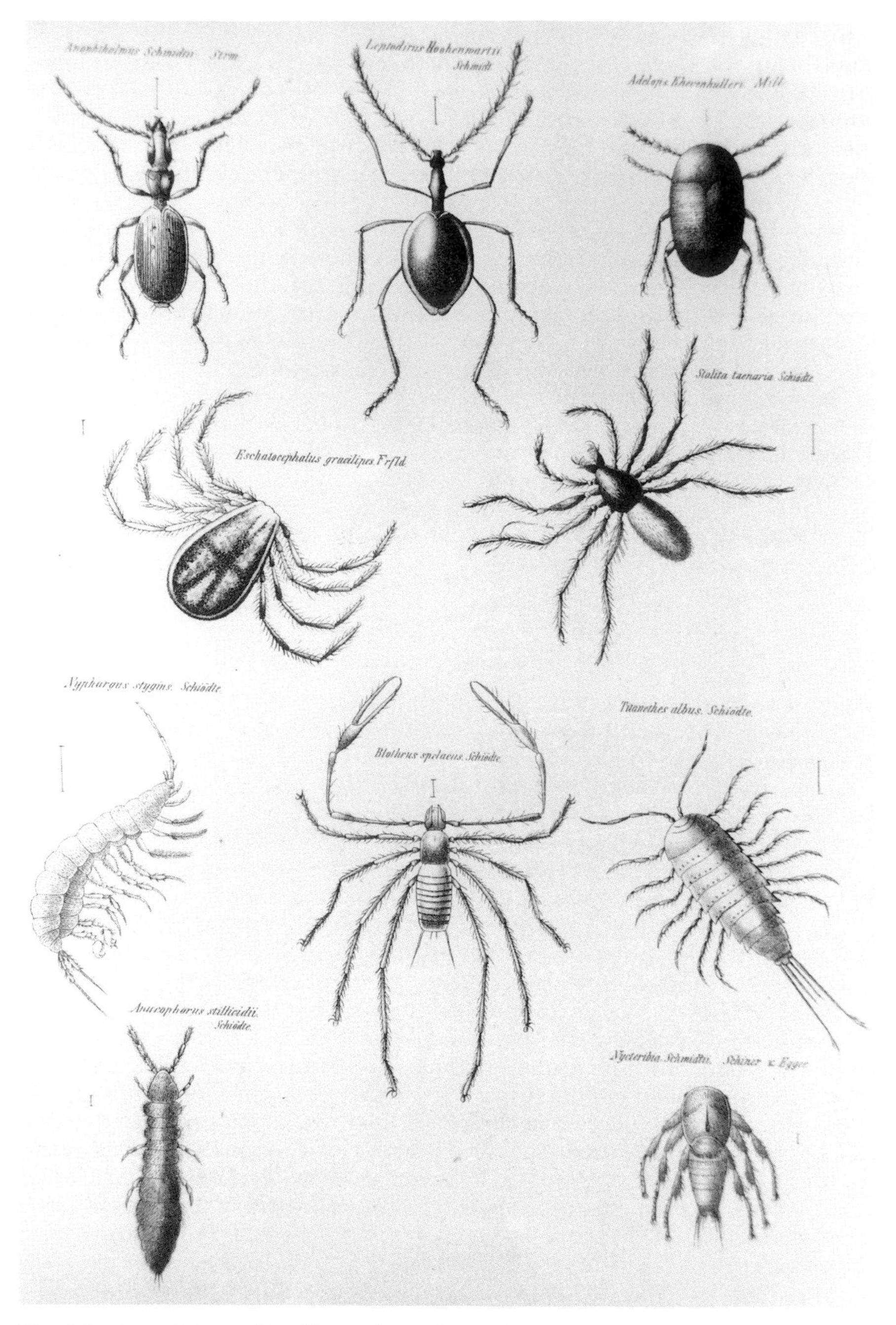

Fig. 1.9 One of the earliest illustrations of cave fauna from Adolf Schmidl's *Die Grotten und Höhlen von Adelsberg, Lueg, Planina und Laas*. Wien, Braumüller, 1854. (Courtesy Trevor Shaw)

In 1936 the British Speleological Association was launched, with a brief to co-ordinate the work of caving clubs and to foster interest in the scientific aspects of caving. Things did not run entirely smoothly, however, and in 1947 another body, the Cave Research Group of Great Britain, emerged with a more specific research interest. Both societies ran in parallel until 1973 when they merged to form the British Cave Research Association which has become a major publisher of speleological research.

Meanwhile another organization concerned with cave science had been formed in 1962. This was the Association of the Pengelly Cave Research Centre, now the William Pengelly Cave Studies Trust. It is London-based, but its interests are very much centred in Devon where it runs the Pengelly Cave Research Centre at Buckfastleigh. The trust is active in education and conservation and produces publications covering a broad range of speleological topics.

The multidisciplinary nature of speleology allows significant contributions to be made as much by talented amateur observers as by trained professional scientists, and we owe much of our present knowledge of the faunas of British caves to the work of a handful of exceptionally dedicated amateur naturalists. The central figures of the group were Brigadier E.A. (Aubrey) Glennie and his niece Mary Hazelton, who in 1938 began making systematic collections in the caves of Yorkshire, Derbyshire and Mendip. Glennie, an excellent all-round naturalist, picked up his interest in cave life while serving in India, where among other things, he published a study on the nesting behaviour of Himalayan Swiftlets in caves. On his retirement in 1946, he became a driving force in the biological work of the newly formed Cave Research Group, and was soon recognized as an authority on British *hypogean* amphipods. Hazelton assumed the mantle of Biological Recorder to the CRG, and for the next 29 years diligently co-ordinated the identification of collections submitted by fellow cavers and compiled the results for publication, first in the Transactions of the Cave Research Group and later of the British Cave Research Association. Among the most notable contributors to the faunal collections of this period were Jean Dixon of the Northern Cavern & Mine Research Society and W.G.R. Maxwell of Chelsea Speleological Society.

The 1950s saw the appearance on the scene of two particularly influential figures, both professional biologists. One was Dr Anne Mason-Williams, a microbiologist whose pioneering studies on the microflora of South Wales caves remains the definitive work in this field. The other was Dr G.T. 'Jeff' Jefferson, a lecturer in zoology at University College, Cardiff, who quickly established himself as the leading authority on British cave faunas and went on to become president of the British Cave Research Association, and a greatly respected ambassador for speleology in Britain. Jefferson's major contribution to cave science in Britain before his untimely death in 1986, was in shaping the wealth of observation gathered by his amateur predecessors into a coherent picture of the biogeographical history and ecological relationships of our cave fauna. It is his work above all that has provided the inspiration for this book.

Non-cavers are fond of asking cavers why they venture underground. The usual answer is along the lines that "caving is good fun". Many would add that caving is most fun when spiced with the excitement of discovery. For the sporting caver, this means finding a way into previously unvisited passages, or whole new cave systems. For the speleologist there is the further excitement of recording new observations and of gaining fresh insights into the history, development, or life of the cave. The discipline of cave biology remains poorly developed in Britain and Ireland, affording tremendous scope for discoveries of all sorts by amateur as well as professional naturalists.

Driving curiosity and a sense of wonder are perhaps the two features which above all unite the caver and the naturalist. I hope that this book can make the passion of the one intelligible to the other, and so enhance the experience of both.

2

The Cave Habitat

What is a cave?

Put this question to any wetsuit-clad, hard-hatted individual found walking across the Mendip Hills, Yorkshire Dales or the shining limestone pavements of the Burren, and you will discover that a cave is a naturally-formed hole in limestone which is large enough to be explored by a caver.

Ask the same caver what he or she has noticed in the way of living creatures in caves, and the answer may well be "not a lot." It will perhaps surprise most cavers (and naturalists) to learn that over a hundred species of invertebrate animal have been recorded as maintaining permanent populations in cave habitats in Britain and Ireland, plus another score or so species of creatures such as bats and moths which use caves as a regular part-time shelter. The cryptic community to which these creatures belong remains largely undetected by cavers because it generally avoids the relatively large tunnels and booming chambers, the glistening calcite draperies and crashing waterfalls which so captivate the human cave enthusiast, preferring instead the cosy confines of smaller cracks and crevices. I hope in this book to shed light on at least a portion of this unsuspected world which lies beneath our feet, whether we live in Grassington or Glasgow, Lisdoonvarna or London, and to offer pointers to fellow amateur naturalists towards fruitful areas of investigation for the future.

To unravel the natural history of a cave, or indeed any habitat, we must try to perceive it as far as possible from the point of view of its inhabitants. It requires a considerable effort of imagination for such sight-dependent creatures as ourselves to grasp the essence of life in the dark and labyrinthine realm of the cavernicole. A little inspired speculation may be needed to find the 'right' questions to lead us to fresh insights into the mysterious world beneath us.

We might start by imagining what kinds of habitats could be accessible to the cavernicole and then consider which environmental characteristics of such places are likely to influence its choice of where to live. Almost at once we run into problems, for although a good deal is known about the environment in man-sized, air-filled limestone caves in Britain and Ireland, we know much less about the conditions within smaller cracks and crevices or in caves beneath the water-table, and less still about our submarine caves. Fortunately, such information is more readily available from other parts of the world. So in this chapter we will take an international approach to defining and classifying the cave environment, before turning in later chapters to a detailed consideration of what is known about our own cave fauna.

What then is a 'cave' as perceived by its inhabitants? The dictionary definition of "a natural underground chamber" gives us a less than helpful starting point, for why should we suppose that the cavernicole will distinguish between natural or man-made tunnels, or between subterranean and above-ground

Fig. 2.1 The main tunnel of Sleets Gill Cave in Wharfedale, Yorkshire Pennines – a classic phreatic tube, formed and enlarged by water filling the passage and so dissolving the limestone rock equally on all sides. (Chris Howes)

enclosed chambers, as long as the appropriate conditions of food supply and microclimate are present in the living space? Should we then include mines, adits, buried pipes, culverts, sewers, cellars, tombs, the London Underground System, or perhaps even houses and other enclosed buildings in our preliminary list of potential cave-habitats? To what extent should our definition specify the material bounding the cavity? Must a cave be rock-lined, or should we widen our brief to include animal burrows and other spaces present in soil, for surely it must be arbitrary to distinguish between an earthy burrow and a muddy hole of similar dimensions in rock?

As it happens, soil faunas are very well documented, and while it seems that many animal groups are common to both soils and caves (and indeed to leaf-litter and the deep moss-carpets of tropical regions as well), the fauna of organically-rich topsoils is sufficiently distinct from that of most rock-space habitats to warrant a separate treatment. I shall therefore exclude soils forthwith from our definition of cave habitats (but see 'Cave sediments' under 'Types of cave habitat' later in this chapter). Similarly, the voids in other organic, living or once-living materials, such as wood or the guts or blood vessels of animals, have distinctive specialized faunas of their own, clearly distinguishable from those of habitats within inorganic materials – although some specialized xylophages, such as termites, and endoparasites, such as tapeworms or flukes, share certain morphological specializations (eyelessness, depigmentation) characteristic of the more specialized cavernicoles. When it comes to holes of human fabrication, most significant biological criteria must lead us to include them in our category of caves. That they have a very poor fauna in comparison with natural caves, is due less to their artificial nature than to their frequent isolation from sources of natural colonization and their often unfavourable microclimate.

Having narrowed our definition of the cave environment to 'habitable voids bounded by walls of rock, or similar inorganic materials', let us now consider the physical criteria which may determine their habitability: the presence or absence of light, physical space (the size of the hole), the medium filling the space (water or gas mix), the microclimate within the medium (the pattern of change in temperature, pH, etc. over time), and the nature and amount of available food.

Let us begin with the business of light, a variable of obvious biological significance. Beyond the limits of light penetration, the cavernicole will be obliged to rely on senses other than sight, and on foods other than green plants. Perpetual darkness is a characteristic of most rock void habitats anyway, so let us choose to define 'the cave' as a habitat entirely without natural illumination. This will substantially simplify our task, by excluding from the cave fauna a whole host of organisms which seek shelter in cave entrances, but also live in a wide range of other shady, sheltered habitats such as the woodland floor, or river gorges, or houses and other structures used by people. Later we will consider the illuminated portions of man-sized caves as a significant 'cave-related habitat' – the 'cave threshold' – simply because it is familiar and accessible to cavers, while ignoring all other lit, cave-related habitats.

In the world of dark holes, the physical dimensions of a potential habitat are of obvious importance in determining what creatures can colonize it. One has only to consider the relative body-diameters of a man (say 450 mm across the shoulders), a Greater Horseshoe Bat (60 mm), a cave spider such as *Meta menardi* (6 mm), a springtail (0.6 mm) or a nematode worm (0.06 mm) to appreciate that one creature's spacious accommodation may be another's unenterable squeeze, and that the cave biologist may be excluded physically from all but a tiny proportion of the very largest of cave habitats. Frank

Howarth, an entomologist who works mainly on the fauna of Hawaiian lava caves, distinguishes three principal hole-size categories which appear to have biological significance for subterranean biotas. He terms these 'macrocavernous' (>200 mm diameter), 'mesocavernous' (1–200 mm diameter) and 'microcavernous' (<1mm diameter).

The characteristic inhabitants of Howarth's microcaverns are sometimes termed 'the interstitial fauna'. They include a distinctive suite of specialized, skinny-bodied crustacea (such as *Bathynella* and various harpacticoid copepods) and other tiny creatures (such as rotifers, nematodes and tardigrades) which mostly like to be in contact with a solid surface on all sides and typically inhabit the spaces in between unconsolidated, fine-grained sediments such as the sand and gravel of river beds and the seashore. I propose, on purely arbitrary grounds, to exclude this fauna from further discussion in this book (except for species which also frequently inhabit larger spaces), and to restrict the definition of the cave habitat to holes of 1 mm diameter upwards, that is, to mesocavernous and macrocavernous habitats.

Various vertebrates use macrocavernous caves (and the larger mesocaverns) for shelter and they, and the other species which depend on their presence, form characteristic communities which reach astonishing levels of diversity and abundance in tropical regions. I shall long remember my first visit to the spectacular Deer Cave in the Gunung Mulu National Park in Sarawak, where at dusk close on half a million bats stream out of the cave in a seemingly-endless cloud which winds its way across the sky with a rush of wings like the sound of Niagara Falls. British bat-watchers have to be content with the odd flap, but in spite of declining populations, cave-roosting bats are still widespread and bat caves do support their own suite of associated 'batellite' cavernicoles. Other cavernicoles may, for example, be specifically associated with the guano of cave-roosting crickets, or with cave sediments introduced by sinking streams.

Mesocavern-sized holes not only occur within karstic rocks, but also in screes, in the coarse gravels and rocky beds of upland rivers, between the pebbles and cobbles of exposed sea-shores, in the fractured zone of non-karstic rocks (especially shales) just beneath the soil, and as cooling cracks in lava flows and other igneous rocks. They represent a very much larger habitable subterranean space than do macrocaverns and so have developed a richer and often more specialized fauna, frequently dominated by species peculiar to this habitat and characterized by a reduction in the size of the eyes, loss of pigment and various other specializations. These 'mesocavernicoles' may also occur in soil spaces, or animal burrows, or even in large macrocaverns, provided there is an adequate food supply of down-washed organic material and a fairly stable humid microclimate. Not all species within the mesocavernous fauna will be found in all related habitats; some do not seem to occur in soil-spaces, others shun human-sized caves.

Simply as a consequence of our own species' enormous body-size, we are physically excluded from the very habitats which are most likely to harbour a specialized fauna. In the absence of appropriate tools with which to peer inside mesocavernous habitats, cave biologists have so far been forced to infer what they can about them from the behaviour of their biotas where they pop up in the accessible portions of people-sized caves. These act as windows into the mesocavernous world, but it seems likely that they provide a distorted view, encouraging widely differing interpretations of the nature of what has been observed. The present situation in cave biology is a bit like that which prevailed among astronomers a century or so ago, when dependence on inadequate earth-based optical telescopes sustained the widely-held belief that Mars was criss-crossed by an elaborate network of irrigation canals built by Martians. Speculation and controversy abound no less in cave biology literature, while cavernicolous communities remain enigmatic and under-recorded. As a result, new species await discovery in most subterranean habitats in every part of the world including the British Isles. In short the whole subject of cave biology is very much still in its infancy. A nice illustration of this turned up on my desk in the form of a report from Frank Howarth, announcing his discovery of a brand-new diverse fauna of highly specialized cavernicoles in lava caves in tropical Australia. For years Australia was thought to have a very poor fauna of specialized cavernicoles, and a number of papers sought to explain this on theoretical grounds. For example, it was argued that Australia's climate during the Pleistocene had not been harsh enough to exterminate the above-ground populations of its cavernicoles, and so any tendency on their part towards specialization for underground life would be continually cancelled out by gene flow from outside the cave. Having wickedly sub-titled his paper *Why there are so many troglobites* [= highly specialized cavernicoles] *in Australia*, Howarth makes the telling point that "One has to actually enter a cave and *look* for troglobites before proclaiming on theoretical grounds that none could exist." I offer this creed to the reader in the context of the British cave fauna. Let us, as naturalists, devise ways to find out what lives in our underground world and get down and study it at first hand.

I have distinguished between 'interstitial' (microcavernous) and 'cave' (meso- and macrocavernous) habitats on the grounds that their biotas are substantially distinct. We might expect a similar distinction to exist between 'aquatic' and 'terrestrial' cave communities. Certainly, there are some cavernicoles which are essentially aquatic, and others which are essentially terrestrial. However, the atmospheres of most mesocavernous, and of some macrocavernous, gas-filled habitats are permanently saturated with water vapour. This poses physiological problems for many groups of terrestrial arthropods which, unless equipped to eliminate excess water from their tissues (as aquatic species do), would quickly die of 'water poisoning' through dilution of their body fluids. Not surprisingly, 'terrestrial' mesocavernicoles have been found to be physiologically specialized to cope with a hydrating atmosphere and seem able to withstand long periods of immersion in freshwater – an adaptation which is essential in habitats which are frequently flooded by downward-percolating rainwater or by fluctuations in the water-table. Some seem equally at home in air or water, and can frequently be seen feeding on the floor of cave pools among their aquatic counterparts. Conversely, many freshwater aquatic mesocavernicoles seem able to cope with 'terrestrial' life without undue physiological stress and have been recorded as living out of water for several weeks at a time. So we see that the distinction between terrestrial and freshwater aquatic cave habitats is not exactly cut-and-dried, although there is a clear distinction between the communities present in either zone and those found in marine cave habitats.

Not all terrestrial cave habitats are moist. Large caves with more than one entrance often experience drying airflows which can produce desert-like conditions which are lethal to the hygrophilic denizens of the mesocaverns. However, such caves are often easily accessible and attractive to vertebrates, and may (especially in the tropics) support vast populations of bats, birds and guano-associated invertebrates. Guanobious animals exhibit few or none of the morphological characteristics considered by European cave biologists to be the mark of a 'true cavernicole' or 'troglobite', yet they may be just as exclusively cave-dwelling as any mesocavern specialist. Above-ground human structures are usually designed to be as dry as possible and are seldom completely dark, and this makes them suitable as a habitat for only a very few cave-threshold specialists, such as the daddy long-legs spider *Pholcus phalangioides* which presumably originated somewhere in the Mediterranean region, but in the UK is found only in houses. In between the dry, draughty macrocaverns and the soggy, airless mesocaverns, there may be wide expanses of transitional cave habitats with a variable microclimate, posing a distinct set of problems for the communities which inhabit them. Terrestrial inhabitants of such places must cope with the physiological stress of desiccation some of the time and physiological drowning for the rest of the time. In the tropics, transitional cave habitats may be particularly extensive, with their own specialized faunas, often dominated by 'bandits' – marauding predators and scavengers which live off the scraps of the guano-based community. Climatically similar conditions are found in man-made culverts and other artificial tunnels, and these frequently attract transitional-zone cavernicoles such as the widely distributed cave spider *Meta menardi*. Later we shall distinguish a range of natural and artificial cave habitats principally on the basis of their microclimatic regimes.

In earlier discussing the criteria which may be important to cavernicoles in choosing their habitats, I included the apparently pedantic phrase 'gas mix', rather than 'air' in my list of the media which may fill mesocavernous voids. I did so because it seems that the atmosphere of mesocaverns may differ substantially from that found in open macrocaverns with a good air circulation (which generally have much the same atmosphere as the outside world). Bacterial decomposition of organic material in small spaces frequently results in unusually high atmospheric concentrations of carbon dioxide. Frank Howarth's new Australian cavernicoles, mentioned earlier, were found in poorly ventilated lava caves which are thought to share the atmosphere of the mesocavernous spaces in the surrounding basalt. The air in these caves is saturated with water vapour and contains around 250 times more carbon dioxide than normal air. Bad air caves occur in Britain too, but have not yet been biologically investigated.

Finally, we may seek to distinguish cave habitats from non-cave habitats in terms of their food supply. Early cave biologists, whose experience of cave faunas was mainly confined to the larger, more easily explored 'fossil' macrocaverns (those no longer bearing the watercourses which formed them) of temperate European limestone areas, concluded that cave animals were perpetually starved. While food resources may be very thinly distributed in such cave habitats, in others (and particularly in the tropics) food may be superabundant. The biotas of food-poor caves are adapted to eke out what little energy is available, while those of food-rich caves are adapted to a life of plenty. Caves may contain a wide range of food sources, including living vegetation (tree roots, saprophytic plants and fungi which get their energy by digesting organic matter rather than by trapping sunlight, fruits carried in by vertebrates), living invertebrate or vertebrate animals, and all kinds of detritus. Cavernicoles may be plant-, fungus-, detritus- or bacteria-feeders, predators, parasites or a combination of these. In short, caves are more ecologically diverse than most biologists realize.

To summarize then, cave habitats may be defined as **'perpetually-dark voids, more than one millimetre in diameter (and sometimes much larger), bounded by rock or similar inorganic materials, and filled with gas ('fresh' or 'bad' air) and fresh or salt water.'** Within such habitats, the microclimatic regime and the type and quantity of the available food-supply largely determines the species composition of the cave community. Only the largest (and often, in our islands, the least populated) cave habitats are accessible to human observers, so that we know a good deal less about the composition and functioning of cave communities in Britain and Ireland than we do about most other natural communities of our islands.

What lives in caves?

Of the voluminous literature dealing with the biota of caves, two works of this century stand out for sheer scope of vision. The first, B. Wolf's *Animalium Cavernarum Catalogus*, published in three parts between 1934 and 1938, lists all animal species recorded from caves to that date. The second, by A. Vandel, published in French in 1964, discusses the biota and biology of caves worldwide. An English translation, published by Pergamon Press in 1965 as *Biospeleology: The Biology of Cavernicolous Animals*, is perhaps still the most useful general text despite its wacky view of evolution in caves. L. Botosaneanu's book *Stygofauna Mundi*, published in 1986, gives a more up-to-date account of the fauna of subterranean waters, but there is need for a similar treatment of the terrestrial cave fauna to take into account the spate of biological discoveries in tropical caves during the decade and a half since Vandel's book. The following brief summary illustrates the range of life forms presently known to inhabit caves.

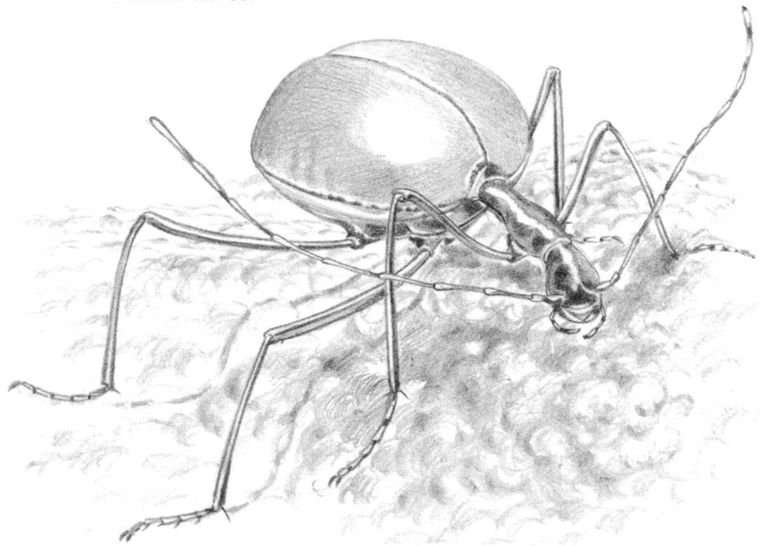

Fig. 2.2 Leptodirus hohenwarti, a highly cave-evolved beetle discovered in 1832 by the Count von Hohenwart in the Slovenian cave of Postojna Jama.

Kingdom MONERA

Phylum Bacteria
Well represented in caves. Includes saprophytes, pathogens and chemo-autotrophs (which live by oxidizing or reducing iron and sulphur compounds). Bacteria are at the base of many cave food-chains.
Phylum Cyanobacteria
Some species are capable of synthesizing their pigments in the absence of light. Various Chroococcaceae are implemented in the formation of complex cave mineral deposits such as *moonmilk* and *tufa* (see glossary).

Kingdom PROTISTA

Phyla Phytoflagellata, Zooflagellata, Sarcodina, Ciliophora, Sporozoa
Protista are often abundant in interstitial waters and many species occur in caves. In Turkmenistan, brackish wells in the Kara-Kum desert contain abundant populations of at least 10 species of unusually tiny, thin-shelled Foraminifera. Cave clays often contain Mastigophora, Sarcodina, Amoebina and some Ciliata.

Kingdom PLANTAE

Phylum Chlorophyta
Various free-living algae, such as *Chlorella, Scenedesmus and Pleurococcus* are found growing deep inside caves. Though able to synthesize pigments in the dark, they appear to use non-photosynthetic metabolic pathways. The other plant phyla Rhodophyta, Phaeophyta, Bryophyta are essentially absent from the dark parts of caves. The phylum Tracheophyta is represented by a very few aberrant saprophytic species which can live independently of sunlight.

Kingdom FUNGI

Phyla Zygomycetes, Ascomycetes, Basidiomycetes, Myxomycetes
Fungi are important in cave ecosystems. Most are saprophytic on organic material washed into caves and form the main food base of cave communities. Some are epizoic (live on the outer surface of animals) or parasitic. Most members of the phylum Oomycetes are parasitic on flowering plants and therefore not represented in caves.

Kingdom ANIMALIA

Phylum Porifera
Encrusting sponges may be the commonest organisms in tidally flushed submarine caves.
Phylum Coelenterata
Hydra viridissima occurs in *groundwaters* in the Southern Carpathians of Europe. Marine Hydrozoa and Anthozoa are commonly found in submarine caves.
Phylum Platyhelminthes
Rhabdocoel Turbellaria are common in wells, springs and groundwaters. Cave-evolved Triclads have a worldwide distribution, with most species within three planarian families: Dendrocoelidae, Kenkiidae and Planariidae.
Phylum Nematoda
Free-living nematodes are frequent in groundwaters, caves and mines worldwide. Several freshwater species of the otherwise exclusively marine Desmoscolecidae inhabit caves in Slovenia.

Phyla Nemertinea and Rotifera
A few species of these small creatures inhabit interstitial waters and caves.
Phylum Annelida
Submarine caves often contain huge populations of sedentary polychaete worms and a number of cave-evolved freshwater polychaetes are known from Switzerland, Slovenia, Japan and Papua New Guinea. Oligochaete worms are often abundant in caves, in groundwaters and sediments. The family Lumbriculidae contains many essentially cavernicolous species. Cavernicolous leeches are known from Central Europe and several tropical countries.
Phylum Mollusca
There are many cavernicolous Gastropoda within the families Auriculidae, Zonitidae, Subulinidae, Enidae and Valloniidae (terrestrial species), and Hydrobiidae (aquatic species). The Bivalvia include cavernicolous species of *Pisidium* and *Congeria* in Europe and Japan.
Phylum Onychophora
The South African species *Peripatopsis alba* is known only from caves.
Phylum Arthropoda
Subphylum Crustacea
Class Remipedia: Recently discovered in submarine caves in the Bahamas and Canary Islands, these actively swimming predators resemble aquatic centipedes.
Class Ostracoda: There are many groundwater species, some of which also occur in larger caves.
Class Copepoda: Many cyclopoids and harpacticoids occur in caves and interstitial groundwaters worldwide.
Class Malacostraca: The monospecific order Spelaeogriphacea is so far known only from caves in South Africa. Thermosbaenacea and Mysidacea are widespread in submarine caves and brackish groundwaters and a few species have made the evolutionary shift into freshwater caves. Isopoda are one of the best-represented orders in the subterranean world, with hundreds of cavernicoles described within the Asellidae and Oniscoidea. The Amphipoda, notably the Gammaridae, are equally well represented. Cavernicolous decapods such as crayfishes, galatheids, crabs and river prawns are particularly widespread in the tropics. Groundwater-inhabiting members of this class often belong to ancient lineages and the distributions of related taxa provide important evidence in reconstructing the history of the planet.
Subphylum Uniramia
Class Diplopoda and Pauropoda: Worldwide there are hundreds of species of cavernicolous millipedes, and particularly of Polydesmoidea. They often show *relictual distributions* which mirror past configurations of the earth's crustal plates, long since redistributed by seafloor spreading and continental drift. A few Pauropoda are known from caves.
*Class*es Chilopoda and Symphyla: Cavernicolous Scolopendromorphs and Lithobiomorphs are known from European caves. In the tropics there are also cave-evolved Scutigerids. The Symphyla live in the soil and look like cave animals, but most feed on living plant roots and so are excluded from deep cave habitats.
Class Insecta: The Collembola and Diplura are primitive, wingless, ground-dwelling insects which require a high humidity. They include a large number of cave-evolved species worldwide. Cavernicolous Blattodea are mostly confined to the tropics. Within the Orthoptera, the Rhaphidophoridae or camel

crickets contain a number of conspicuous cavernicoles with a worldwide distribution and a few cavernicolous gryllids are found in the tropics. There are cave-evolved Dermaptera, Psocoptera, Hemiptera, Trichoptera, Lepidoptera and Diptera. Not surprisingly, the majority of cavernicolous insects belong to the largest order and the most successful group of animals on earth, the beetles (Coleoptera). At least 22 families contain cavernicoles, which reach their greatest diversity and specialization in the Trechidae and Leiodidae.

Subphylum Chelicerata

Class Arachnida: A few cavernicolous Scorpiones, Uropygi, Amblypygi, Schizomida, Ricinulei and Palpigradi are known from tropical regions. Pseudoscorpions are well represented in caves, with over 300 cavernicolous species, many of which are giants among their kind, with very long legs and claws. The Opiliones also contain cavernicoles, most notably in the families Phalangodidae, Travuniidae and Ischyropsalidae. The terrestrial mites (Acari) most frequently found in caves are tiny Gamasides within the families Parasitidae, Rhagidiidae and Eupodidae and there are many cavernicolous water mites in the families Hydrachnellae and Porohalacaridae. Spiders (Araneae) dominate the predator niches in most tropical caves and include hundreds of cavernicolous species worldwide. In temperate regions, the most specialized cavernicoles belong to the primitive families Dysderidae, Leptonetidae, Telemidae and Oonopidae, but there are also many specialized species of Linyphiidae, Erigonidae and Agelenidae.

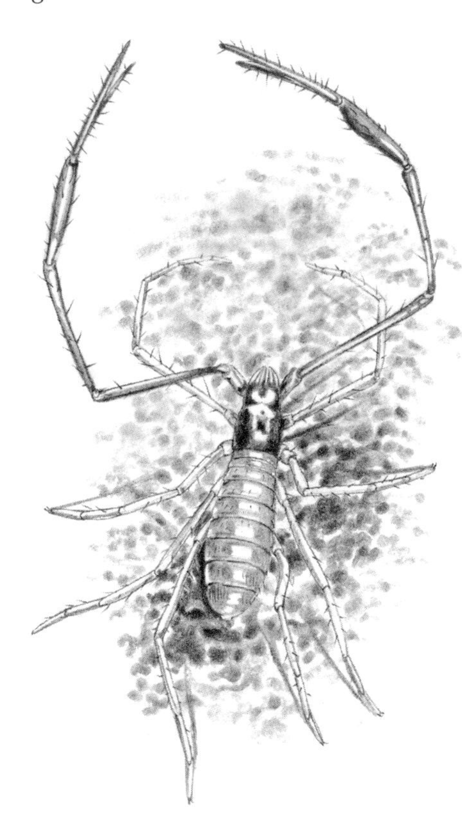

Fig. 2.3 *Neobisium spelaeum*, a giant cavernicolous pseudoscorpion which preys on the cave beetle *Leptodirus hohenwarti* (shown in *Fig. 2.2*)

Phyla Tardigrada and Bryozoa
A few species inhabit groundwaters.
Phylum Echinodermata
Various detritivorous brittle stars (Ophiuroidea) and sea cucumbers (Holothuroidea) frequently occur in submarine caves, but no strictly cavernicolous species are known.
Phylum Chordata
Class Teleostomi: Freshwater cavernicolous fishes are found mainly in the more arid regions of the world. To date around 60 more or less blind and depigmented species representing 8 orders and 13 families have been collected in freshwater and submarine caves, springs and groundwaters. The most speciose families are Cyprinidae, Gobiidae and Bythitidae, followed by Pimelodidae, Characidae, Cobitidae and Amblyopsidae.
Class Amphibia: 14 species of cavernicolous Urodela are known from North American caves and one from Europe. The latter, *Proteus anguinus* was the first cavernicole to be recognized as such. In common with species of the American genera *Gyrinophilus, Eurycea, Typhlomolge* and *Haideotriton, Proteus* retains larval gills in the adult stage, and is thus able to live permanently beneath the water table.
Class Reptilia: The colubrid snake *Elaphe taeniura* is a common inhabitant of caves in south-east Asia, from China to Borneo, where it preys on bats and swiftlets.
Class Aves: While no birds live permanently in caves, swiftlets of the genus *Aerodramus* (Apodidae) and the oilbird *Steatornis caripensis* (Caprimulgidae) are dependent on caves as nesting sites and are capable of navigating long distances underground and in total darkness by *echolocation* using *ultrasonic* clicks.
Class Mammalia: It is possible that a South American mouse *Heteromys anomalus* and one or two species of tropical shrews in the genus *Crocidura* may establish permanent cave populations. Many bats (e.g. Tadarida spp.) are dependent on caves for shelter, but none are cavernicolous.

In summary, although green plants are largely absent from caves, cavernicolous species are found within most of the major classes of animals, and are particularly common among the Crustacea, insects, spiders and millipedes.

Caves in limestone

Of several cavernous rocks in Britain and Ireland, one above all others provides extensive integrated cave systems of a size which allows the biologist to study the life they contain with relative ease. It is the Carboniferous Limestone.

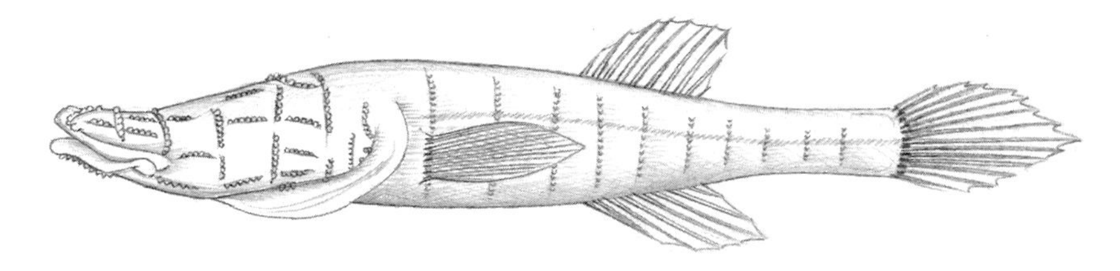

Fig. 2.4 The blind North American cave fish *Speleoplatyrhinus poulsoni* has an extraordinarily developed lateral line system, with correspondingly enlarged brain parts for processing tactile and positional information.

The cave-bearing limestones are concentrated in six major patches: in County Fermanagh (the largest, but least well prospected area), the Northern Pennines, Derbyshire's Peak District, County Clare, Somerset's Mendip Hills and South Wales. In the Northern Pennines alone, there are 1800 or more documented cave entrances and around 350 km of explored and mapped cave passages – while current estimates give the total length of mapped passages in Britain and Ireland at somewhere around 800 km. Cave exploration is still in a very active phase in these islands, and significant new discoveries are still being made. For example, recent explorations in Ogof Daren Cilau beneath Mynydd Llangattwg in South Wales have yielded over 20 km of new discoveries in less than two years, including the largest passages yet found in Britain.

By any standards, 800 kilometers of open cave passage represents a significant habitat, worthy of the attention of naturalists – yet passages of explorable size must form but a tiny proportion of the total cave habitat, the vast majority being of mesocavernous dimensions. In the absence of data, I would guess that the habitable surface area within the mesocaverns of limestone terrains must run to at least two or three orders of magnitude more than that within explored caves.

So what is it which makes limestones so spectacularly cavernous? To understand the process in which caves are formed, we must begin by examining the origins, nature and structure of the limestone rock itself. Limestone is a sedimentary rock, that is, it began life as suspended particles in an ancient sea, gradually settling to the ocean floor millimetre by millimetre over millions of years. During this inconceivably long period, there were intervals when conditions changed enough to interrupt the steady downward rain of lime, allowing some other type of deposit to intervene briefly in the sequence of otherwise pure calcium carbonate. Aeons later, and now hardened to rock, these geological glitches have become 'fossilized' as *bedding planes* – horizons of weakness between the solid layers of limestone.

The Carboniferous Limestones of the British Isles were laid down somewhere in the tropical seas of the southern hemisphere. Pushed along on a northward-drifting chunk of continental crust, they have had a bumpy ride. Some, like those of the Yorksire Dales, have survived their 340 million year journey the right way up, though somewhat fragmented by massive vertical faults. Others have fared less well. The Mendip limestones lie like a wrecked car, buckled and perched at a steep angle, so that the bedding planes dip downhill at an average gradient of 50° or so. In all cases, the rough ride has produced vertical stress cracks, called *joints*, which link with *faults* and with the original bedding weaknesses of the limestone to form a boxwork of crevices reaching from the highest hilltop to beneath the deepest valley.

Limestone is a strong rock and so frequently forms upland regions. Solid limestone is impervious to water, but water is able to flow through the cracks within it. It is these cracks which are the key to understanding the origins of caves. Limestone caves form principally by means of a simple chemical reaction in which hydrogen ions from groundwaters, acidified with dissolved carbon dioxide, act on the relatively insoluble carbonate ions in the limestone to produce soluble bicarbonate ions which are then flushed away. The reaction renders the limestone 25 times more soluble than it would be in pure water and the result is holes.

Some of the carbon dioxide in groundwaters is collected by raindrops falling through the atmosphere, and some from the breakdown of organic material picked up as the rain then trickles down through the soil. Immediately beneath the soil, the weathered surface layers of rock are more fractured than those at a greater depth and the acidified, aggressive soil waters have their maximum impact here – so that at any one time, up to 15% of the volume in the top three metres of limestone may be occupied by air-, or water-filled spaces (Stearns, 1977 – figures for Central Tennessee, USA). These mesocaverns act like the guttering beneath the roof of a house – collecting soil water and quickly conveying it to natural drainpipes, often developed on the intersections of major joint fractures, or steeply inclined bedding planes.

The flow of water downwards into the limestone carries with it sediments which contain particulate organic material, dissolved organic acids and micro-organisms (bacteria and fungi). Decomposition continues way below the soil in the cracks and crevices of the limestone. There, to paraphrase Hoover's famous advertising slogan, groundwater micro-organisms 'eat-as-they-seep-as-they-clean', mopping up the organic impurities and excreting CO_2 – in effect arming their liquid medium with chemical teeth. In the larger conduits it may take up to 50 days and several kilometres of flow before the bacteria finish mopping up the water-borne food, and even longer and further before the chemical aggression of the cave water is finally spent. All this time, limestone is being steadily corroded, and the cracks along which the water travels widened, resulting in the slow opening of a complex drainage network reaching deep below the water table.

The initial pattern of flow within the flooded cracks is dictated by the structural geology of the limestone and the shape of the land surface. Between them, these two factors determine where water will escape from the rocks as a spring, and lay down the blueprint for the caves to come. If the rock strata lie horizontally, as they do in the Yorkshire Dales, water is forced to follow a rectangular course down vertical joints and along short sections of horizontal bedding, producing the characteristic stepped profile of a 'pothole' system. If the strata are inclined, as in the Mendip Hills, drainage will alternate between short joint-controlled shafts and longer bedding-plane slopes, producing caves with a steep profile, but few vertical sections.

At or below the water table, the course taken by percolation water is determined by the direction of the hydrological gradient between where the water goes into the permanently-flooded system of cracks known as the 'phreas' and where it comes out again as a spring. Within the phreas, water is free to follow along, down or up the 3-D maze of cracks to produce the smoothest possible overall flow. Where the rock beds lie horizontally the smoothest profile may be along just one bed, producing a horizontal cave which may run for several kilometers (Yorkshire has many such systems). Where the rocks dip steeply in a direction which does not coincide with the hydrologically-determined direction of flow, the groundwater may be forced into a series of vertical z-bends, down joints and up the bedding, to tack its way to the spring (a classic example is Wookey Hole in the Mendip Hills). Where the limestone beds are trapped in a syncline or U-bend beneath impermeable rocks, water may be forced to travel down to great depths following the configuration of the rock strata. Chinese geomorphologist Yuan Daoxian has recently reported the discovery of a substantial cave at a depth of 2900 m below the water table in the Sichuan

Basin in China. When perforated by drilling, the hole gushed water. Caves at this depth are, however, extremely rare and probably have little or no biological significance.

As the profile of the cave takes shape, under the twin controls of *hydrology* and geology, the 'best route' is inevitably favoured with a greater rate of flow, which promotes more rapid corrosion of its boundary walls – which in turn results in a still greater flow capacity. In this way, the initially diffuse drainage within the limestone is gradually simplified with increasing time (and depth) into a pattern of coalescing collectors of increasing diameter. Over the aeons, the vagaries of geology and hydrology may conspire to favour one particular route in preference over all others, opening it out to form a major trunk conduit which drains the entire *subterranean aquifer* to its spring, or 'resurgence'.

I have suggested so far that cracks in the limestone can be enlarged to form mesocavernous conduits by corrosive water trickling gently down towards the water table under the influence of gravity, or by slow creeping but aggressive groundwaters draining towards a resurgence. However, chemical solution is not the full story – once the cave becomes large enough to be an efficient drain, fast-moving waters can carry sharp-edged grit, adding physical claws to chemical teeth. Grains of quartz (a mineral much harder than calcite) provide a particularly effective scouring agent, and are easily collected by streams running over the Millstone Grit which butts onto much of our cavernous limestone. Where such a stream is captured by a well-developed Yorkshire joint, the outcome is a vertical *pothole*; on Mendip, an inclined *swallet cave* results.

Erosion of the river valleys which drain a limestone block may lower the water table within it, so that a conduit formed within the flooded *phreas* eventually becomes emptied of water and filled with air. One such system, GB Cave in the Mendip Hills, has received intensive study in recent years – in turn by Drs Tim Atkinson, Pete Smart and Hans Friederich of Bristol University. Their work has demonstrated the extent and importance of the system of mesocavernous conduits which overlies and feeds water and sediment into GB Cave. The conduits can be divided into three types. First there is the dense network of '*subcutaneous*' cracks in the top three metres of rock beneath the soil. These have a large total volume and drain rapidly into the underlying cave, via mesocavernous collectors. They operate intermittently for up to two days following rain, but do not store a great deal of water. Two other types of inlets feed into the cave. They are small-diameter seeps, which stay full of water all the time and show little variation in flow discharge (flow being constrained by their small diameter); and *vadose* flows, which have higher and more variable discharges than do the seeps. Smart and Friedrich report that inlets are more frequent into shallower cave passages, and less frequent at depth, but do not give estimates of the porosity represented by these inlets, for which we have to consult Stearns on his Tennessee karst. Thus, at 10 m depth, the initially high porosity within the zone of subcutaneous flow (15%) may have fallen to about 1.5%, with a mainly downward flow via mesocavernous shafts and vadose cracks. At 30 m depth, porosity is down to 0.002%, and flow (at least in massively-bedded limestones) is increasingly confined within macrocavernous collector passages, fed by a converging network of smaller collectors of mesocavernous size. Thus, the main concentration of cave habitats (of mesocavernous dimensions) lies within the top three metres of bedrock.

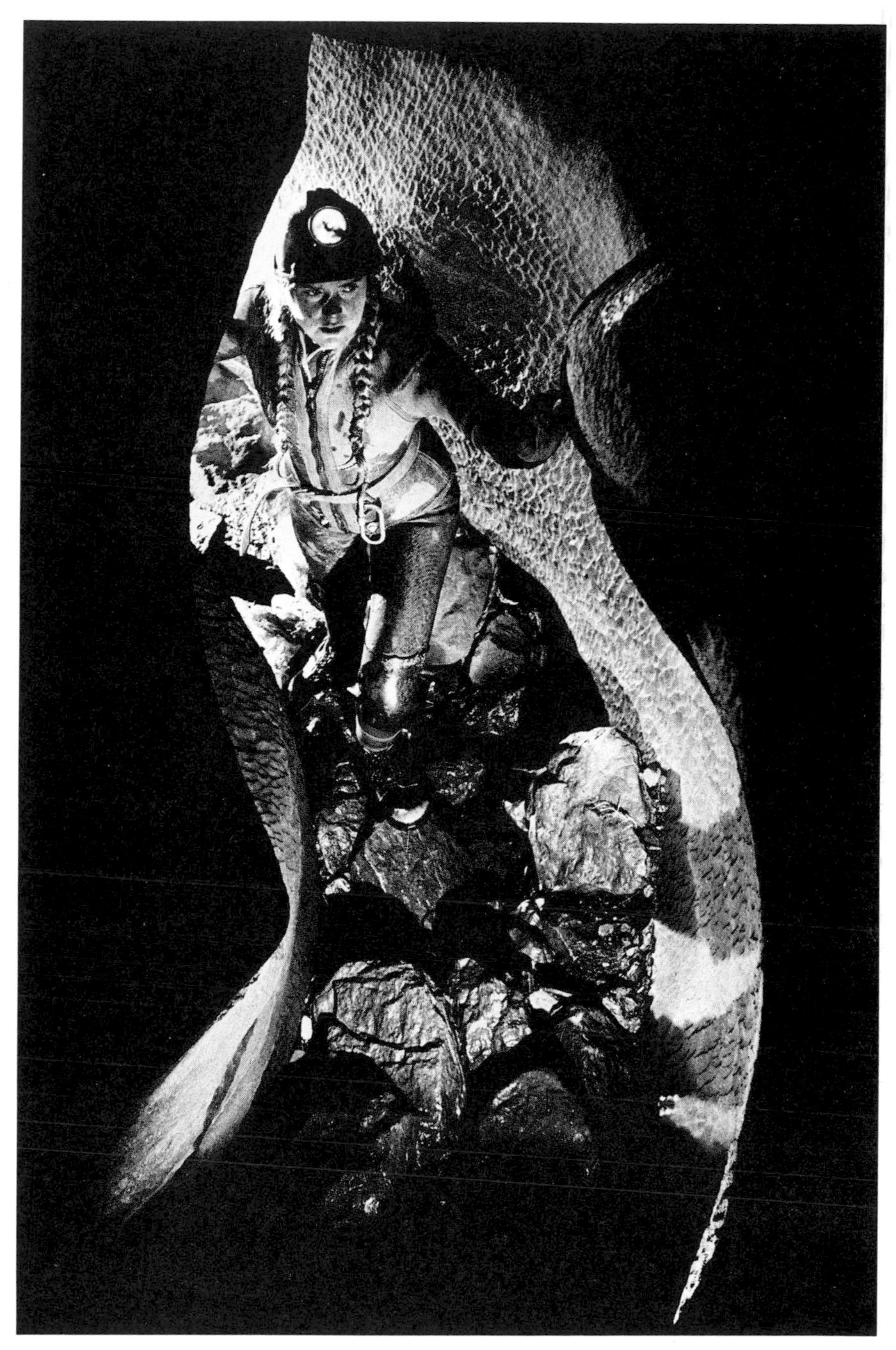

Fig. 2.5 The main streamway in Poulnagollum Pot, Co. Clare – a classic vadose passage cut down through the limestone beds by an allogenic stream flowing off the eastern flank of Slieve Elva (compare with *Fig. 2.1*). (Chris Howes)

Phreatically-formed passages can be recognized by their characteristic circular or at least symmetrical cross-section, produced by equal *corrosion* of the walls, ceiling and floor (a classic example is Peak Cavern in Derbyshire). As soon as they develop an air surface, corrosion and *erosion* become channelled downwards and the cave stream begins to cut a trench into the floor, as is seen in the Ogof Ffynnon Ddu streamway in South Wales. This may eventually result in a canyon-shaped passage many metres deep.

As the cave matures, the enlargement of joints and bedding planes may weaken whole blocks of the ceiling and floor, resulting in their collapse. Where the cave contains an *aggressive* stream, such material may be removed as it falls, creating a chamber whose size is limited only by the structural strength of the overlying rock. Given enough time, the roof will eventually fail, opening the cave to the sky; if the collapsed blocks are not removed by water, the cave fills up with rubble.

The same chemical process which resulted in the removal by solution of limestone in cracks close to the surface, can go into reverse when acidified, lime-laden waters drip through cracks into an underground passage full of fresh air. As CO_2 in the hanging water droplet diffuses out into the air of the cave, the resulting transformation of bicarbonate to carbonate ions forces lime to precipitate from solution. The drip falls, a rim of lime remains on the cave ceiling, and hey presto, a few thousand years later the cave is festooned with sparkling *stalactites*. Given an aeon more, these may completely fill up and block the passage.

Caves may also become filled with insoluble sediments from outside. This is frequently the case in Britain, where sediments enter the cave by one of three main mechanisms: large masses of unsorted clay-and-rubble have slumped into caves as a half-frozen mush during the glacial advances of the Pleistocene; sands and gravels are washed in by cave streams and dumped as the waters slow down in more gently graded stretches of cave; and fine mud is often trickled downwards by *percolation* water until it completely fills passages right up to the roof.

The worst ravages of collapse and sedimentation may be reversed by a little-known bio-chemical phenomenon known as 'digging'. First noted in Mendip caves, it appears to be caused by large sweaty men in overalls, armed with plastic buckets, and generally takes place at weekends in the pursuit of new explorations and discoveries underground. The chemical component of the process consists in a sparing application of 'Dr Nobel's Linctus' to otherwise immovable blockages, with sometimes impressive results. We shall consider the conservation implications of digging and blasting in caves in a later chapter.

Types of cave habitat

In the previous section, we traced the process of cave formation in limestones by the gradual chemical enlargement of bedding cracks and joints, their subsequent drainage, further enlargement, collapse and infilling by sediment, breakdown and *speleothems*. In the course of this 'life cycle', limestone caves present a series of distinct habitats which are each exploited by a characteristic biota.

In recent years, similar biotas have been found in equivalent habitats in a wide range of other rocks (lava, gypsum, mudstone, shale, chert, breccia, tuff, rhyolite, diorite, quartzitic sandstone, etc) and sediments. This has led to a plethora of exotic technical terms in the literature – it seems that any biologist

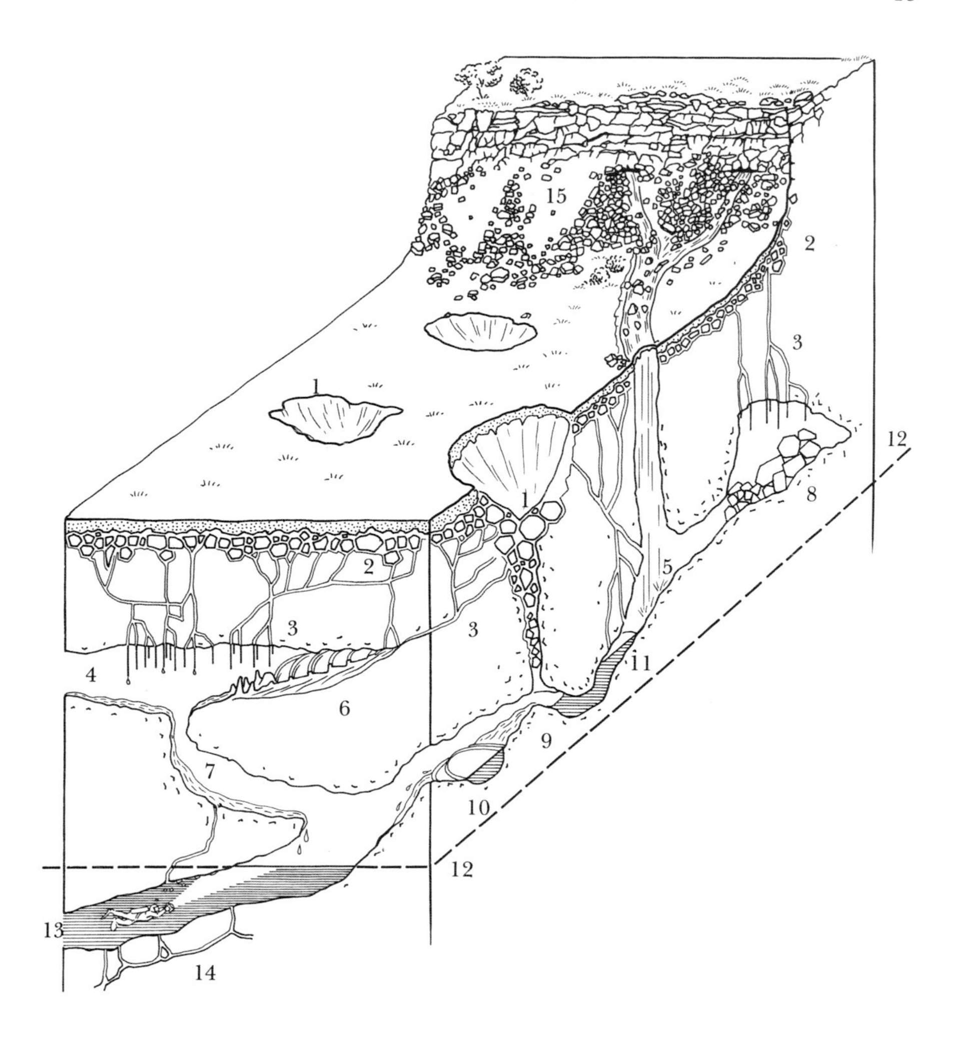

Fig. 2.6 Cave habitats and structural features in a limestone upland.

1. Dolines or shakeholes 2. SUC or subcutaneous zone 3. Mesocavernous seeps and conduits 4. Air-filled macrocavern or dry cave passage 5. Vadose macrocavern or wet cave passage 6. Rimstone gour pools 7. Cave sediment 8. Breakdown blocks 9,10. Riffle and pool in cave system 11. Perched siphon or sump 12. Water table 13. Water-filled macrocavern or sump in the phreas below water table 14. Water-filled mesocaverns in the phreas 15. Talus, below a scar or limestone cliff.

who finds a new habitat wants to give it an impressive, polysllabic label, an urge doubtless born of years of grappling with latin species names. My favourite examples include 'parafluvial nappes' (as worn by water-babies?), 'hypotelminorheic biotope' and 'the petrimadicolous biocoenose' (etymologically-inclined readers may enjoy trying to sort out these little beauties). In the section which follows, I have attempted to keep things simple by stressing the the similarities between habitats, rather than the differences between them.

Air-filled macrocaverns (dry caves)

The longest, biggest and, by any criteria, the best macrocaverns are those developed in limestone. The world's longest cave is the Mammoth Cave system in Kentucky at over 540 kilometres, while the deepest is the Réseau Jean Bernard in the French Pyrenees at over 1600 m deep. Our longest cave, the Ease Gill system in Yorkshire, 70 km long, can manage only 15th place in the world ratings. The ages of caves vary considerably. Recent estimates suggest that certain limestone cave passages in North America could have formed 10 million years ago, while some tropical limestone caves are estimated to be only around 100,000 years old.

Substantial caves are not by any means confined to limestone. Other notable caves are formed by solution of marble, dolomite and non-carbonate rocks such as gypsum (e.g. Optimisticheskaja in the Ukraine, 183 km long); rock salt (in arid regions); and even quartzitic sandstone (e.g. Sima Aonda in Venezuela, 362 m deep) – while lava tubes are formed by the crusting-over and subsequent draining of molten lava flows (e.g. Kazumura Cave in Hawaii, over 20 km long). The latter include the youngest of all caves, formed on the slopes of Kilauea Volcano in Hawaii within the last five years. Other significant caves may be formed in a range of different rocks and within accumulations of large talus boulders (e.g. Lost Creek in Colorado, USA), by gravity sliding (gull caves), tectonic movements, wind-, or wave-blasting and other mechanisms. Most bizarre of all is Kitum Cave on Kenya's Mount Elgon, which has been literally *eaten* for 200 m horizontally into a bed of volcanic ash by generations of salt-hungry elephants.

In caving parlance, a 'dry cave' is one which a human visitor can explore without getting wet. While such caves may provide an easy route into an underground world of great beauty, budding cave biologists may be greatly

Fig. 2.7 Mexican Free-tailed Bats, *Tadarida brasiliensis,* stream out of the entrance of Carlsbad Caverns at dusk. (Chris Howes)

disappointed not to find every surface festooned with strange, eyeless arthropods, armed with sweeping antennae, stalking around on matchstick legs. As stated earlier in this chapter, dry passages in tropical caves may contain huge populations of bats or birds and a wealth of guano-associated cavernicoles, but in Britain and Ireland, cave-roosting bats and the species which depend on their presence have become increasingly scarce in recent years. Nevertheless a surprisingly large number of our macrocaverns do still harbour small populations of bats and a few associated 'battelite' species, which will be considered in Chapters 4 and 5.

The main biological interest of our dry caves is that they may contain the only accessible populations of cavernicoles primarily adapted to other, relatively inaccessible habitats. Thus, drip- or seepage-fed pools, *gour pools* and wet *flowstone* may harbour aquatic mesocavernicoles; and rock piles, flowstone pockets and 'deep cave' or 'stagnant air' environments may give us a rare glimpse of the terrestrial mesocavernicolous fauna.

Vadose macrocaverns (wet caves)

Cave streams may contain water derived from permanently-flowing surface streams, swallowed through an open sinkhole, or percolation water which has collected gradually from a large number of mesocavernous inputs. Those fed entirely by percolation water or seasonal run-off tend to hold more biological interest, as the populations they contain (being genetically isolated from surface habitats) may have evolved specialized features.

Surface streams arrive in the cave complete with a biota of their own, most of which will survive quite well in darkness for some considerable time, though few species may succeed in reproducing. Apart from the lack of growing green plants, and a tendency to experience less seasonal temperature change, such streams do not differ significantly as a habitat from streams above ground. The terrestrial habitat in stream passages is usually draughty and moist. Its fauna may include the adult stages of aquatic insect larvae swept in with the sinking stream, and various predators, such as spiders, which feed on them, plus a range of detritivorous cavernicoles.

Where stream passages enter the phreas (cavers' 'sumps', marking the end of exploration to all but divers) it is not unusual to find accumulations of organic detritus deposited by floods. Given a suitable microclimate, such places may be well populated by specialized cavernicoles, or 'troglobites'. Indeed, in many food-poor cave systems (such as those in mountainous areas of the Pyrenees and Cantabrians) this may be the only habitat in which cavernicoles occur in any numbers.

Water-filled macrocaverns (phreatic caves)

The chemistry, food supply and fauna of phreatic waters is determined to a large extent by their mode of entry into the cave. Underground waters are classed as autogenic (originating as through-soil percolation via diffuse seeps and subcutaneous flow) or allogenic (sinking streams). The former tend to be lower in organic content, less chemically aggressive by the time it reaches the phreas and may carry a lower sediment load. Phreatic waters may have a very different chemistry to the vadose waters which feed them, because they are unable to de-gas the carbon dioxide produced by microbial oxidation of organic materials, or to replenish oxygen used up in such decomposition.

The rate of flow in large-diameter phreatic tubes is generally greater than in phreatic mesocaverns, but is still often sluggish enough to accommodate small, slow-swimming cavernicoles (such as *Niphargus fontanus*), which avoid fast-moving vadose streamways.

In recent years cave divers have penetrated great distances into freshwater phreatic macrocaverns, and to considerable depths, but to date no detailed studies have been made of the biota of this remote and fascinating environment.

Water-filled mesocaverns

Most of our knowledge about the structure of mesocaverns comes from looking at exposed joints or bedding planes in limestone quarries, or occasionally caves, and from the work of karst hydrologists. We know that the very earliest stages of cave development occur under phreatic conditions, eventually creating inter-connected systems of conduits which may be in any plane, from horizontal to vertical. Phreatic mesocaverns may take the form of wiggly networks of small-diameter tubes (*anastomoses*), or thin, but laterally-extensive cracks, or narrow shafts – and are probably as common in other cavernous rocks, such as chalk or gypsum, as in limestone. Evidence that such spaces are inhabited comes from the animals found in well-water over the centuries. It is ironic that the earliest records of our cave fauna should be from a habitat about which little more is known today than a century-and-a-half ago, when Philip Henry Gosse wrote:

> "recently, investigations in various parts of the world have revealed the curious circumstance of somewhat extensive series of animals inhabiting gloomy caves and deep wells, and perfectly deprived even of the vestiges of eyes ... even in this country we possess at least four species of minute shrimps [all of which] have been obtained from pumps and wells in the southern counties of England, at a depth of thirty or forty feet from the surface of the earth."

There is no way at present of collecting information directly about how aquatic cavernicoles use mesocavernous bedrock cracks, but it seems likely that great local variation exists within this habitat in terms of oxygen concentrations, pH and food supply. Such factors are likely to influence the distributions of the fauna, and will be discussed in Chapter 4. Fortunately there are other, more accessible types of water-filled mesocavernous habitat which are easier to study. They include the deeper interstices of stream-bed cobbles (*phreatic nappes*), and a peculiar sub-soil phreas of mesocavernous dimensions which occurs on the surface of impervious silt or clay deposits in mountainous areas of Europe (the *hypotelminorheic medium*). In both these habitats, the food supply comprises dissolved or finely particulate organic material, and the waters tend to be rather low in oxygen; and both contain faunas very similar to those of limestone mesocaverns.

Amphibious mesocaverns

As no detailed investigation of the biology of air-filled mesocavernous habitats has yet been attempted, we are forced to infer what we can about the conditions within them from studies of limestone caves and other similar habitats.

As soon as mesocaverns develop an airspace, they become available for colonization by terrestrial cavernicoles. However, cracks and anastomoses are extremely flood-prone, often filling up with water each time there is heavy rainfall at the surface. Vertical cracks probably flush more violently, but remain water-filled for shorter periods than horizontal cracks, and this may result in some differences in their faunas. Less immersion-tolerant organisms may tend to inhabit the wider, better-drained vertical cracks and humid terrestrial cave habitats, while the more aquatic organisms may prefer horizontal cracks or cave pools. It is likely that particulate organic material accumulates at specific points within the cracks (perhaps at the upstream ends of permanently flooded sections), so that some patches of habitat will be better supplied with food than others. Some areas may be too anoxic to support any life other than anaerobic micro-organisms, while some patches may harbour relatively large concentrations of detritivorous invertebrates. As previously discussed, there may be an almost complete overlap in distribution between the 'terrestrial' and 'aquatic' components of the fauna of such habitats.

The French biospeleologists Juberthie, Delay and Bouillon consider that mesocavernous spaces in fractured rock immediately below the soil constitute a habitat which is separate from caves, which they have termed the 'Superficial Underground Compartment' (SUC). Their claim rests on differences between the fauna found here and that found in deep caves. They consider the primary cause of such differences to be the greater temperature variation experienced in the 'SUC' compared with the 'Deep Underground Compartment' (DUC) represented by deep fissures and caves. While this may be so in SUC habitats beneath shallow soils of regions which experience a strongly seasonal temperate climate, I would doubt that the microclimate in deeply-buried SUC habitats or those of tropical karst differs a great deal from that of the 'DUC' – and there is evidence that cave faunas migrate up into the SUC periodically in order to exploit the resources they contain. For the purposes of this classification, I propose therefore to treat the 'SUC' as part and parcel of other intermittently-flooding mesocavernous spaces, whether they be immediately below the soil, within cave passages, or connecting one with the other.

There would seem to be little doubt that the SUC within calcareous rocks is by far the most extensive and important of all cave habitats in terms of the numbers and diversity of its biota. Since their 'discovery' of the 'SUC' (an environment previously well-known to karst hydrologists as the 'subcutaneous zone'), Juberthie and Delay have gone on to show that this habitat and its biota not only occurs in limestone and other cavernous rocks, but also in 'non-cavernous' shales, granites, schist, gneiss, sandstones, etc. My first reaction on reading the paper announcing this discovery was to attack the bottom end of my garden with a pick and shovel. There, to my delight and amazement, I found tiny-eyed cave spiders (*Porrhomma egeria*) frolicking among the fractured chunks of Pennant Sandstone just one metre beneath the wreckage of the flower bed. As far as I know there has been no systematic investigation to date of the fauna of 'SUC' habitats in Britain and Ireland – an extraordinary gap in our knowledge which surely must be remedied before long.

A better-known mesocavernous habitat is contained in talus, or scree, whose surface can frequently become covered with vegetation and soil, turning it into a fair imitation of Juberthie and Delay's SUC. When not sealed by soil, the upper levels of talus are unsuitable as a habitat for cavernicoles, being too cold in winter, too hot in summer and too dry for much of the time. However, if the scree is deep enough, the lower levels must surely provide exactly the conditions favoured by cavernicoles, though I know of no work on this deep-talus habitat in Britain.

I know of only two accessible 'DUC' mesocavernous habitats within caves. One is in the spaces within rock piles (underground talus), the other is in speleothem pockets. Rock piles may, or may not provide a suitable habitat for mesocavernicoles. If the pile is in an old, dry 'fossil' passage, as most rock piles tend to be, it is unlikely to contain enough food to support life (unless the cave contains bats, or other vertebrates, in which case the rocks may be over-run by guano-beasts). On the other hand, if the pile is sufficiently extensive, and is traversed by percolation water carrying organic material, it is likely to harbour a rich fauna of mesocavernicoles – although the depth within it at which a searching biologist can expect to 'strike bugs' will increase with the increasing dryness or breeziness of the surrounding cave atmosphere, precisely as would be the case with above-ground talus. Juberthie (1983) gives an interesting example of a schist-boulder pile in the great Salle de la Verna chamber in France's Pierre Saint Martin cave. It is inhabited by a typical SUC fauna of *Aphaenops* beetles which appear to be quite oblivious of the fact that their schist scree habitat lies the best part of 1000 m underground.

Speleothem pockets are essentially just spaces of mesocavernous dimensions like all the others described in this section, but where they occur in the 'deep cave' or 'stagnant air' zone of caves (see the microclimate section, later in this chapter), they will often prove to be the very best places to look for mesocavernicoles. Speleothems occur where percolation water, rich in dissolved lime, intersects a cave passage. Where such deposits are laid down over mud, pockets often form between the two; and if deposition is still in progress, trickling water maintains just the microclimate conditions favoured by cavernicoles, while also supplying a source of food. In short, they perfectly reflect conditions in the mesocaverns. The late A. Vandel, in his famous book *Biospeologie la biologie des animaux cavernicoles* (published in 1964), described such a habitat in the Grotte de Sainte-Catherine, at Balaguères, in the Ariège region of France:

"One side of the chamber is formed by a stalagmitic wall which is covered by a thin layer of water which flows from an opening in the roof ... The constant flow brings in organic material from the exterior which nourishes the Collembola and nematoceran Diptera on which *Aphaenops* (a blind cave beetle) feeds ... The stalagmitic covering is separated from the wall by a space of a few millimetres which contains the products of dissolution of the rock: clotted red and black clay, and black magnesium deposits.
When one enters this chamber three or four *Aphaenops* can usually be seen running on the wall in search of food ... If they are caught, others appear. Closer examination soon explains this phenomenon. The calcareous wall is formed of stalagmitic columns arranged parallel to each other. Between these columns small holes have been hollowed and it is into these that *Aphaenops* disappears ... It appears that the space between the rock and the stalagmitic wall constitutes the biotope in which Aphaenops lives when not in search of food, and in which they reproduce."

Cave pools

Gours, or rimstone pools are formed by seep-fed, lime-rich waters trickling down an incline in an open cave and depositing a sequence of curved, retaining dams. Such pools presumably provide conditions akin to those in phreatic mesocaverns, and often harbour a similar fauna, which is augmented in the cave by animals washed out from seepage cracks during heavy rain. Other drip-fed pools may also contain aquatic mesocavernicoles, providing they receive a food supply.

The concave meniscus of pool surfaces may act as a deadly trap for soil and mesocavernous springtails (see Chapter 5), providing a happy hunting-ground for other specialized mites and springtails, whose feet are equipped to grip.

Cave sediments

All kinds of sediments find their way into caves. Many arrive complete with their own specialist biotas: bacteria, fungi, nematodes, earthworms and so on, and these may be exploited by cavernicoles as a source of food. Cave sediments may also serve as a habitat for the eggs, larvae, or pupae of cavernicoles.

Moonmilk

Moonmilk is a term applied to white, wet, cheese-like or dry, sticky, powdery formless masses found in limestone caves. This weird stuff may contain a cocktail of carbonate minerals – including calcite, aragonite, monohydrocalcite, magnesite, hydromagnesite, nesquehonite and huntite – some of which are alledged to be associated with particular bacteria isolated from moonmilk. A blue-green alga *Synechococcus elongatus* found growing in moonmilk in complete darkness must have been using alternative metabolic pathways to the normal photosynthetic ones it employs in the light, and other algae (*Gleocapsa magna*) may also be present. I know of no cavernicolous animals associated with moonmilk, so this is not a habitat which I propose to discuss further in this book.

Submarine and intertidal cave habitats

The submarine Green Holes and intertidal Brown Holes of Doolin, in County Clare, are ordinary limestone caves, formed in the usual way, which became inundated by rising sea levels at the end of the last glacial period – about 12,000 BP. No doubt they have had a long history of intermittent sub-aerial cave-development during cold glacial periods of low sea-stance, alternating with periods of immersion in sea-water during interglacial warm spells, such as the planet currently enjoys. They provide a significant habitat for marine life – Mermaid's Hole has been explored for over a kilometre and the Hell complex totals several hundred metres. Other significant submarine caves occur in the Brixham area of Devon and at Durness in the far north of Scotland. Otter Hole, a resurgence cave which opens on to the banks of the river Wye near Chepstow, has a tidally-flushed entrance series which may contain fresh or saline water, depending on the height of the tide and the volume of the cave river. The faunas of such caves will be discussed in Chapter 5. In other countries, an interesting fauna has been found in coastal brackish groundwaters, but I know of no such British fauna. It may exist, but no studies have been made.

Submarine caves, such as the Green Holes, should not be confused with 'sea caves', such as the famous Fingal's Cave in the Hebrides, which are spray-filled holes blasted out of cliffs by wave action. The latter may harbour a few ubiquitous, fast-moving cliff-dwellers such as sea slaters (*Ligia oceanica*) and silverfish (*Petrobius maritimus*), which can dive into cracks to escape the force of the waves, as well as a few of the more hardy marine organisms found in the more extensive drowned limestone caves referred to previously. In short, the fauna of sea caves is unremarkable.

Slutch caves

Slutch is the onomatopoeic term given by Wainwright to the peat bogs of Kinder Scout, Bleaklow and Black Hill in Derbyshire. Apparently, water flowing between the base of the peat and the underlying gritstone has carved out a number of caves of explorable dimensions – one of which has been followed for 50 m underground by cavers Steve Fowler and Tony Moult. It seems that the base of the peat is riddled with such caves and with smaller air- or water-filled passages of mesocavernous dimensions. No doubt this will prove to be a widespread habitat in upland peat deposits throughout Britain and Ireland, perhaps with a characteristic fauna all its own, which for the moment appears to have escaped investigation.

Mines, tunnels, cellars and tombs

As artificial caves, such spaces may be inhabited or visited by any cavernicoles which have both the motive and the opportunity to do so. The motive will be a suitable medium/microclimate and food supply (see following sections). The opportunity will fall to any cavernicole whose habitat intersects the artificial cave.

Food supply

Chlorophyll and sunlight are the elements of life on earth and the source of that unique green glow which identifies our living planet when seen from space. Photosynthesis lies at the base of the food chains in which all life is meshed: fish, fowl and fungus; tree, turtle and tiger – or almost all.

In 1977, an American ship on an oceanographic survey south of the Galapagos Islands, located some active volcanic vents on the sea floor at a depth of 3 kilometres. An instrument pod was sent down, armed with video cameras to record the scene. As the probe moved towards the vents, the watching scientists were amazed to see their monitor screens fill with a writhing mass of enormous worms, 10 centimetres thick and up to three metres long. Close by were beds of 30 cm-long clams. Among them swam shoals of fish, and white crabs scuttled across the black basalt rocks. At these depths, far beyond the reach of sunlight, life is generally thin on the ground. A few starfish, crinoids and crustaceans subsist on the steady drizzle of detritus, and are in turn eaten by predatory fish. To find such an abundant local concentration of large organisms clearly pointed to an unusually rich food source. The mystery was soon solved. The volcanic vents, it seems, were spouting superheated, sulphur-laden water. As this cooled, clouds of black sulphides formed and were immediately consumed by great concentrations of bacteria. The worms and the clams were feeding on these bacteria and they in turn supported the scavenging fish and crabs. The bacteria concerned are chemo-autotrophs – that is, they can harness the chemical energy in the volcanic sulphides to power their own vital processes. What is more, this whole process and the mini-ecosystem which revolves around the 'volcanic bacteria' is quite independent of sunlight.

Sulphur bacteria, and relatives which derive energy by reducing ferric iron compounds, are common in caves – in sediment banks as remote from solar rays as are the depths of the deepest ocean trench. The muds where they live are home to nematode worms, known bacteria-feeders, which are hunted by tiny, scurrying beetles. How many visitors, catching sight of a cave beetle, appreciate that they may be watching one of the rarest of all living phenomena – a predator sustained, at least in part, by a food chain independent of sunlight.

The energy source for cave-based chemosynthesis generally originates outside the cave, as Carboniferous Limestone itself contains very little in the way of iron and sulphur minerals. These compounds, and the bacteria which exploit them have usually been washed in as part of the cave's sediment load. But there is at least one energy source which may originate within the fabric of the cave itself. Most limestones contain detectable amounts of organic matter, largely in the form of hydrocarbons. Such material would be of no use to animals directly, but if there are bacteria present which are capable of using it as an energy source, it could be continually liberated into the cave ecosystem at the interface between the cave and the rock, imperceptibly, as the cave is dissolved out by flowing water. Since the organic matter in the limestone will have been derived from organisms present in the seas when the rock was being laid down, its energy content must originally have come from the sun. It is in fact fossil solar energy.

Interesting though they may be as a scientific curiosity, chemo-autotrophic bacteria contribute only slightly to the food base of most cave communities. By far the biggest source of energy is still the sun's rays, but at second-hand, in the form of introduced detritus from surface communities; for no green plants can survive in the absolute dark of the cave.

The absence of green plants in caves led early biospeleologists to the conclusion that cavernicoles are starved animals. In 1886, Packard insisted that the shortage of food available to cave animals is the reason for their small size. While it is self-evident that large animals with a high metabolic rate can have no place in an entirely heterotrophic, food-poor ecosystem, in reality many cave species are actually far larger than their surface relatives. Most recent studies have shown that cave-evolved animals (troglobites) have unusually low metabolic and growth rates and that they save energy in every way possible, by streamlining their movements and by adopting highly efficient foraging and reproductive strategies. These are obvious specializations to cope with a low food supply, and seem to be a fundamental characteristic of cavernicolous evolution at temperate latitudes like ours. Some cave species are remarkable in their metabolic efficiency and consequent ability to tolerate starvation. Gadeau de Kerville (1926) reported that a specimen of the Slovenian cave salamander, or Olm (*Proteus*), had been kept in captivity for fourteen and a half years, and for the last eight of these had received no food. He did not report whether it eventually died of starvation or just plain old age.

Some more recent cave biologists have noticed that captive Olms regularly slough and then eat the mucus layer which, like an extra skin, covers and protects their whole body. Mucus is sticky and microscopic examination has shown that in captive amphibians it becomes encrusted with bacteria, algae and protozoa. So de Kerville's amazing 'non-feeding Olm' may all the time have been sneaking clandestine meals of diatoms-in-slime – not the tastiest of fare, but enough to keep it ticking over. Streams sinking into caves must carry with them a fair load of phytoplankton, and it may be that the Olm's mucus-eating behaviour in captivity has some adaptive significance in the subterranean rivers where it lives.

Fig. 2.8 Olms (*Proteus anguinus*) – blind, depigmented cave salamanders which retain gills and an aquatic lifestyle as adults.

In a closed system, such as our Olm aquarium, the recycling of dissolved nutrients could go on endlessly, providing the Olm did not lock them up as extra Olm tissue. Once equipped with a source of energy – sunlight from the nearest window will do – the tank becomes a self-perpetuating 'mini-ecosystem'. Because caves are perpetually dark, production cannot meet demand and they can therefore never be considered as proper self-sustaining ecosystems. But the waters which form, enlarge, infill and eventually destroy caves almost invariably carry organic compounds – complex chemicals gathered from the soil or the breakdown of animal and plant detritus. Organic chemicals (dissolved in water, or clumped together in big lumps of detritus) fill the role of an energy source in the cave. There has been much speculation by cave biologists over the extent to which dissolved organic substances can be absorbed directly by aquatic cavernicoles. So far the evidence is inconclusive, but there is no doubt that they are captured by microfungi, bacteria and protozoa, which are plentiful at least in some allogenic cave waters. So, one way or another, all organic material entering the cave becomes available to the bottom rung of cave animals – the detritivores which fill the equivalent role to the primary consumers of the sunlit world above.

While the photosynthesising parts of plants have no place in caves, there is no reason why their roots should not penetrate into subterranean voids. Indeed, in the lava tubes which run just beneath the skin of the active volcanoes of Hawaii, tree roots form dense subterranean forests which support a range of sap-sucking bugs, root-chewing caterpillars and their predators; pale-skinned, eyeless relatives of species found in the forest above. Roots seldom penetrate into macrocaverns in Britain, but they may constitute a significant food source in a number of other cave habitats, such as the SUC, culverts and slutch caves. No one seems to have investigated root-associated faunas in such locations, but it would surprise me if such a widespread niche is not occupied by at least one insect specialist.

Second-hand plant material takes many forms: autumn leaves carried on a sinking stream, tree trunks hurled into a pothole by violent floods, or well-rotted humus quietly inched down the cracks of a limestone pavement. Usually by the time such materials reach the depths of the cave where the true cave faunas live, they have been thoroughly pulverised by water and rock, tenderised by bacteria and fungi, and often enough, passed through a series of invertebrate guts.

This was brought home to me during my first visit into the higher levels of GB Cave in the Mendip Hills, several years ago. The upper levels of the cave contain a series of narrow rifts which leak water from the overlying land whenever it rains. I was hunting for tiny cave beasts in these upper levels, and looking out for patches of fresh mud which I guessed would be organically richer and so would contain more life. I soon began to realize that each and every sediment cone emanating from each and every leaky crack was made up of thousands of tiny mud pellets, like scaled-down grains of rice. They were arthropod droppings – hundreds of millions of them, forming a deposit which over the centuries was gradually filling up the upper dry passages, and probably the lower reaches beneath the water table too. Fortunately, most cavers are blissfully unaware of the true nature of the substance which they spend each weekend crawling through and, no doubt, liberally ingesting themselves.

Biologists used to studying ecosystems based on living plants too often treat detritivores – animals which eat dead organic material – as a single dietary category. In the cave ecosystem, all the first-level consumers are 'detritivores' and it is surprising just how many different specializations they manage. This is most clearly seen in *guano* communities, where chains of very specialized organisms use different components of the insect remains which form the bulk of bat droppings. Stuart Hill has studied the ecology of one such community in a bat cave in Trinidad. Guano of the funnel-eared bat, *Natalus tumidirostris* (notorious, incidentally, as a carrier of the human diseases, relapsing fever and blastomycosis), was eaten as soon as it fell by a cockroach, *Eublabarus distanti*, which removed most of its fat and some of its protein (much of the unbound protein having been already stripped out in the gut of the bat). The cockroach droppings, still rich in chitin, were decomposed by a fungus, *Penicillium janthinellum*. This in turn was fed on by the mite *Rostrozetes foveolatus* and various other tiny arthropods. The mites in turn were eaten by other arachnid predators. Similar sorts of food chains probably operate in British bat caves, but the research has not yet been done.

In our present ignorance, we really do not know exactly what many of our British cave detritivores get out of their 'junk food' diets. There must be some degree of specialization, because certain types of materials are largely ignored by certain cave animals, while others attract them strongly.

One way to study dietary specialization is to put down baits in the cave and see what species come to them. The baits should be very small or they will provide an un-natural boost to the numbers of certain species and so upset the balance of the cave community.

I tried a baiting programme in the Ogof Ffynnon Ddu system in Wales, during the course of a wider survey of the fauna. I already knew, from a number of earlier collecting visits, that at least six species of fly were common in the upper level passages (a dung fly, a coffin fly, a winter gnat, and three or more fungus gnats), together with eleven springtail species, a millipede, a woodlouse and various beetles and mites. All were associated with detritus in one form or another, so when I put down small pieces of raw liver or cheese in various choice spots, I expected to gather quite a collection of invertebrates. Over the next couple of weeks I came back regularly to check the baits and was disappointed to find nothing on them. By now the liver was beginning to get a bit unsavoury and the cheese had gone slimy. Two and a half weeks after placement, I found every single liver and cheese bait crawling with the slender maggots of the winter gnat, *Trichocera maculipennis*, but still nothing else came near them. Eventually, the larvae crawled away to pupate, the bacterial smell subsided, and only then did a few *Folsomia* springtails gather in the area. So, far from being unspecialised scavengers, these detritivores showed themselves to be quite a discerning bunch.

One of the key factors in determining how the cave fauna will respond to a particular source of detritus seems to be whether it is first attacked by fungi or bacteria. Many fungi exclude bacteria from their chosen food by secreting antibiotics, and avoid other foods of the wrong pH which are decomposed by bacteria. Sometimes the dominance of either one or the other is determined by the size of the potential food source. In the course of a study of lava cave invertebrates on Kilauea volcano in Hawaii, I put out baits of supermarket white bread, to see what would be attracted. Some baits were in the form of crumbs spread along the rough cave wall, others were big chunks which I kneaded into golf-ball-sized lumps. The crumbs quickly went soggy and grew a thin gelatinous slime of yeast-like fungi which attracted cave-specialised millipedes and crickets; while the chunks became bacterial stink bombs infested by scuttle-fly maggots, but shunned by all the true cave fauna.

Several European biospeleologists have noticed a similar specialization between bacterial- and fungal-feeders. In general it seems that 'surface grazing' arthropods, such as millipedes, isopods and springtails, tend to be fungus-eaters; while 'gulpers' and burrowers, such as fly larvae and earthworms tend to be bacterial feeders. However, even closely related species within the same group can specialize to different foods to avoid competition. An example is found in the springtails *Tomocerus minor* and its congener *T. problematicus*, which co-exist in similar habitats in the Grotte de Sainte-Catherine, in the Ariège region of France. The former mainly munches fungi, while the latter feeds largely on bacteria-rich silt.

In general it seems that microfungi, which occur on a variety of substrates, including wood, animal faeces and plant detritus, form the most important dietary component for most terrestrial cavernicoles. In a study of several Virginia caves, Dickson and Kirk (1976) found that the abundance of the cavernicolous invertebrates was correlated with the abundance of microfungi and with high fungal-bacterial ratios, but not with abundance of bacteria or actinomycetes.

Friedrich, Smart and Hobbs (1982) summarized the literature on bacterial counts for cave sediments and waters. As might be expected, heterotrophic bacteria (which feed on inwashed organic material) are far more numerous than autotrophic bacteria (which utilise the oxidation of inorganic compounds) – but there are very wide variations. Sediments give consistently higher bacterial counts, generally of the order of 10×10^6 g^{-1}, but with large variations either side; while cave waters give very much lower counts, generally of the order of 100 ml^{-1}. These authors' own figures for different types of water inputs in Mendip caves are interesting. Swallets feeding directly into open cave passages give the highest figures (500 ml^{-1}). Percolation recharge which is integrated into mesocavernous fissures and conduits has an intermediate bacterial count (50–300 ml^{-1}), while diffuse flow waters in phreatic microcaverns (sampled via boreholes) give a very low count (2 ml^{-1}). One of their conclusions is that water flow through the limestone aquifers feeding the major springs used as drinking supplies around Mendip does not provide any significant filtration of microbial input. While bacterial counts at the major Mendip risings in general do not give cause for public health concern, this may not be the case in other springs, particularly in tropical countries, used as an untreated drinking supply on the supposition that water which bubbles out of the ground must be pure. Sinkholes are widely used as rubbish dumps, attracting disease-carrying rodents, and the water which enters them may reappear many miles away as a spring, perhaps in a different valley. The late Dr Oliver Lloyd, a well-known Mendip caver, contracted Weil's Disease from *Leptospira* bacteria present in Longwood Cave on Mendip, while several members of expeditions to Borneo in 1980 and 1984 (myself included) contracted a similar form of leptospirosis in the famous Mulu caves, and nearly died as a result.

In slow-moving underground streams with mud bottoms, and in drip- and seep-fed pools isolated from swallet-fed streamways, microfungi are more common than bacteria and are correlated with the abundance of macroscopic cavernicoles. Microfungi and bacteria may be utilized directly by such cavernicoles, or may be eaten by Protozoa, nematodes and other micro-fauna which are in turn eaten by Crustacea (isopods and amphipods) and aquatic insect larvae.

I have tried so far to paint a picture of the cave community as a fairly structured world, where, despite the absence of green plants, the lowly animals at the bottom of the food chain still manage a fair degree of dietary specialisation, preferring some kinds of detritus or living prey over others. In fact, the degree of specialization may be even more pronounced: not just according to the nature of the food source, but even according to the size chunks in which it is packaged. Thomas Poulson and Tom Kane, working in the enormously complex Mammoth Cave system in Kentucky, have found that the terrestrial arthropods in the cave use energy availability as the main basis for dietary specialization. Of the 44 species regularly recorded in their study area, 30 were primarily associated with one food, seven with two, six with three, and two with four of the six natural foods available to them. All the species which used more than one food source picked different stages in the decomposition of each of their foods, or different places along the gradient of food concentration so that they fed at a unique level of energy availability. Two millipedes illustrate how this works. *Scototerpes* feeds on the scattered droppings of cave crickets and to a lesser extent on very dispersed leaf litter in the last stages of decomposition and on veneers of leached organic matter deposited by floods on the cave ceiling and walls. The other common cave millipede, *Antriadesmus* specializes on cricket droppings in areas where they form thick deposits. Not surprisingly, the latter species turns out to have a higher metabolic rate and to lay more eggs than its more frugal cousin.

Poulson and Kane's finding may go some way to explaining the unattractiveness of my cheese and liver baits in OFD. It could be that unusual concentrations of protein-rich food entering an oligotrophic (food-poor) cave, may be shunned by the cave community (except guanobia) as they represent a level of energy availability well above that to which the cavernicoles are adapted. But there may be a simpler explanation. Kathleen Lavoie of the University of Illinois noted that certain cave fungi grew better on large droppings (such as those of pack rats) than on those of smaller species. Rapid colonization by fungal hyphae formed a meshwork barrier which seemed to exclude cave arthropods from this otherwise tasty food resource. So, far from cave arthropods shunning large items of food, it may be that other, more opportunistic forms of life actively keep them at bay.

So far, we have considered only detritus as a food-source. While in sheer numbers, detritus-eaters must of course form the bulk of the cave community, it is predators and omnivores which have the greatest scope for dietary specialization, providing they are able to survive at low population densities. The predators in British caves are unspectacular creatures: mites, spiders, beetles and their larvae, amphipods, caddis larvae and flatworms. Once again, we can only speculate as to the degree of ecological specialization which they enjoy – they have not yet received any detailed study in Britain. The Americans, however, *have* studied their own equivalent species in considerable detail and, as there is no reason to expect cave predators of temperate American caves to behave very differently from our own, their research findings are worth repeating here.

Tom Barr studied six cave-evolved ground beetles which co-exist in Mammoth Cave. One of them, *Neaphaenops tellkampfi*, feeds almost exclusively on the eggs of cave crickets. It finds them by smell, digs them out of the sand in which they were laid, then punctures them with specially-elongated mandibles and sucks out the contents. The other five beetles are all closely related in the genus *Pseudanophthalmus*, and all feed on different prey or in different microhabitats. Even the two near-identical-looking species *P. menetriesii* and *P. striatus* feed quite differently: one hunts for small arthropods in rotting wood, the other digs for tubificid worms in cave silt. Our own cavernicolous carabid, *Trechus micros*, feeds in a similar way. I have watched it work over a tidal mud bank in Otter Hole in the forest of Dean, pocking the surface with hundreds of tiny pits, as it thrust its head repeatedly into the soft sediment in search of prey.

The tidal mud of Otter Hole is an unusual cave sediment which supports an exceptional fauna. Estuarine muds of the Bristol Channel have been estimated to have a productivity four times that of good arable soil. They are rich in organic material and correspondingly smelly. The mud is kept sloshing up and down the upper reaches of the estuary by fierce tides which have an amplitude here of 13 m, one of the greatest tidal ranges anywhere on our side of the Atlantic. During the highest Spring tides, when the cave stream is itself swollen by rains, it is usual for the entrance passages of the cave to flood right up to the roof and to stay flooded for several weeks. Cavers know this, and avoid the cave during this danger period. Eventually the trapped waters drain slowly away, but in the meantime their sediment load has settled out as a rich brown deposit which coats the walls and ceiling of the cave right up to the high-tide mark. The new mud is soon invaded by a wealth of invertebrates, including a millipede, *Brachychaetuma melanops*, and a rove beetle, *Aloconota subgrandis*, found in no other British caves. The richness of the cave fauna is a direct result of the floods.

The importance of seasonal flooding to cave invertebrates has been noted by many biospeleologists. In the food-poor alpine caves of the Pyrenees and Spanish Cantabrians, almost the whole fauna lives in the 'intertidal zone' of flood-prone passages. In the enclosed space of a cave, flooding can take one of two forms: 'flash floods' and 'ponding'. In the former, a sudden rush of water temporarily approaches the carrying capacity of a streamway, sweeping all before it. This is the type of flood dreaded by cavers, and passages which are prone to such events are rigorously avoided whenever there is a risk of heavy rain in the catchment area. 'Ponding' floods happen more gently. They usually occur where collapsed blocks or sediment impede the flow of escaping water, so that any slight increase of input causes a temporary pond or lake to form behind the obstruction. The water may have been flowing quite quickly up to the barrier, but now it slows, depositing its sediment load. As the waters recede, the local invertebrates rush out of hiding to 'beachcomb' for the juciest morsels.

In temperate caves, flooding is a strongly seasonal, and therefore predictable, phenomenon – and many cave species time their cycles of activity and reproduction around it. The advantage of synchronized breeding is obvious in cave species which occur at low densities and in which only a small proportion of females are capable of reproduction in a given year. Tom Poulson, of Yale University, has made a special study of aquatic cave communities in the USA. He explains the complex relationship between cavernicoles and floods as follows:

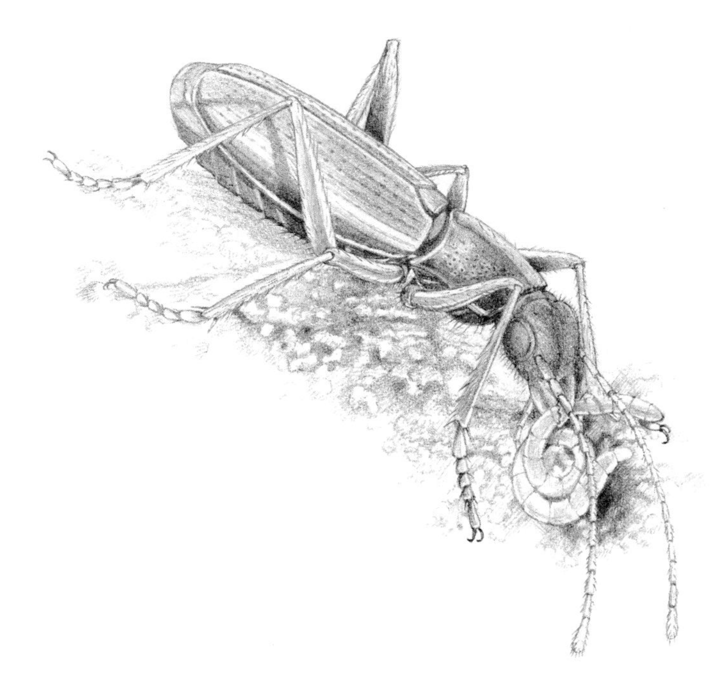

Fig. 2.9 The ground beetle *Trechus micros* digging an enchytraeid worm from the silt of Otter Hole in the Wye Valley.

"Annual growth and breeding cycles in caves are cued by spring floods, specifically by changes in temperature, food supply, amounts of solute, turbidity and current ... Scale growth rings of amblyopsid cave fish, and probably other cave fish, form during floods while plankton and organic matter are being replenished, but the fish are secretive, inactive and not feeding. The amblyopsid cave fish *Chologaster agassizi*, *Typhlichthys*, *Amblyopsis spelaea* and *A. rosae* breed in spring towards the end of the yearly floods or high water, and the young appear during the period of low water 3 to 8 months later when residual food is still present and chances of injury from floods are low. ... Some snails, isopods, amphipods and decapods also breed in spring and early summer ... [for example] breeding in the shrimp *Paleomonetes ganteri* and the crayfish *Orconectes pellucidus* precedes high water, with maximum organic inwash, by 4 to 6 months."

James Keith has found a clear seasonal reproductive cycle in the American cave beetle *Pseudanophthalmus tenuis*, which inhabits flood-prone mud banks in Murray Spring Cave, Indiana. Across the Pacific, Chris Pugsley has studied the New Zealand glow-worm *Arachnocampa luminosa* whose starry displays form the centre-piece of a tourist development at the famous Glow-worm Grotto at Waitomo. Although most stages are present throughout the year, numbers of larvae (the feeding stage) show a clear peak in late spring / early summer, when their food (winged imagines of aquatic insect larvae) are abundant, and the evaporation rate in the cave is at its lowest level.

Seasonal variations in food supply in caves are not due solely to flooding. Bats have a predictable seasonal occurence in temperate caves, and this might be expected to strongly affect members of the guano communities which depend on their presence. Harris (1970) has described the cycle associated with occupation of a cave by a maternity colony of Bent-winged Bats in Australia. As the bats arrive, there are fast changes in food, temperature, moisture relations and pH. Food quantity, rate of daily input and freshness all increase. The guano temperature rises 10°C within a week. The fresh guano, along with bat urination combine to increase the relative humidity from 60 to 95%, and the substrate becomes visibly moist. The urine-ammonia aerosols and faecal material modify the substrate pH and other chemical characteristics. How guanobious cavernicoles respond to such changes does not seem to have been studied in any detail, but it is known that population levels of many species of guanobia increase sharply when fresh guano becomes available, and decrease when it is not.

Microclimate

Until a few years ago, I had often wondered how the 'terrestrial' inhabitants of flood-prone passages and mesocavern cracks survived the regular immersions on which their livelihoods depend. It took a visit to the New Guinea highlands to reveal the secret. I was involved, with a large British expedition, in exploring the huge labyrinth of Selminum Tem – then the longest cave in the southern hemisphere. One day, while a small group of us were in the bowels of the system, the heavens opened and 10 cm of rain fell in a couple of hours. The cave streams rose by several feet in a matter of minutes, and we were lucky to get out in one piece. Two of the team were working in a young, immature network of passages deep below the main trunk of the cave and in their haste to escape the rising water, they dropped a quantity of expensive equipment. So a couple of days later, a colleague and I returned to retrieve it.

The passage had obviously flooded to the roof, and the water level was still falling, amid distant gloops and gurgles. The walls and ceiling of the passage were coated with a thin layer of wet black mud, spangled here and there with fragments of soggy biscuit, washed from a packet dropped by the fleeing cavers two days previously. Several of the fragments had already attracted beetles and millipedes, and as I watched, a glistening wet millipede slowly emerged from the depths of a crack and headed across the mud in the direction of the nearest biscuit fragment. Further along the passage, millipedes of two different species appeared to be feeding on the floor of a temporary puddle -- underwater. Later I watched a woodlouse doing the same thing. It seemed that the cave community here was quite amphibious; sheltering in cracks and crevices as the waters swept through their home and sallying forth to feed once the flood had passed by.

Of course, such behaviour was by no means unknown. As long ago as 1953, those pioneers of cave biology in Britain, the late Brigadier Glennie and his niece, Mary Hazelton had reported:

"We have seen the staphyline beetle, *Ancyrophorus* [now *Ochthephilus*] *aureus* in Bagshawe Cavern, Derbyshire, enter a pool without hesitation and walk across its floor under about 4 inches of water apparently without discomfort ... Owing to the constant high humidity, the distinction between air and water is apparently less readily perceived. Just as the beetle was ready to enter the water, so aquatic crustacea (*Niphargus* and *Asellus* species) are equally prone to leave the water and wander over the moist rocks or stalagmite, finding their way to isolated pools. Terrestrial and aquatic fauna may sometimes be found resting together under a loose stone or other debris on a wet floor."

Fig. 2.10 A millipede *Polymicrodon polydesmoides* emerging from a water-filled crack after a flood. Many terrestrial cavernicoles are adapted to withstand periods of immersion.

More recently, I have watched the little pink woodlouse *Androniscus dentiger* apparently feeding on the floor of a cave pool in Wales, side by side with its aquatic cousin, *Asellus cavaticus*. It would seem that under certain regimes of microclimate, the distinction between aquatic and terrestrial life in caves all but disappears.

Our understanding of the biological significance of cave microclimate has advanced by leaps and bounds in the last few years, largely through the work of Frank Howarth in his Hawaiian lava tubes. Typically two metres or so in diameter, the tubes are the drained-out arteries of spent lava flows. They may run for great distances (20 km or more) at a depth of 5–10 m below the ground and are frequently perforated by skylight vents formed by degassing of the molten rock. Superb examples abound on the flanks of the active Kilauea volcano on Hawaii Big Island, and new ones are formed at each eruption. Lava tube walls are criss-crossed by innumerable cooling cracks which in turn link up with gas-bubble vesicles within the basalt to form a vast coherent system of mesocaverns. Howarth first visited these caves in the early 1970s and was astonished to find a whole community of previously unknown arthropods living on, or associated with, tree roots which had penetrated down the cooling cracks in search of water, and which hung like living curtains within the caves. Intriguingly, many of these cavernicoles could be recognized as close relatives of extant, above-ground species, yet they were clearly modified for their cave existence, with tiny eyes, reduced pigmentation and specialized behaviour patterns. Why had these 'troglobites' become specialized, while many other far older tropical caves were populated by species which showed almost no morphological distinction from surface-dwelling relatives? Howarth hypothesized that some aspect of the lava cave environment might be so different from the cavernicoles' ancestral environments that they had been forced into a complete genetic reorganization in a relatively short period.

In 1980, having spent several years collecting data in the lava caves, Howarth published his results in the journal *Evolution*. The key to understanding the distributions, behaviour and evolution of Hawaiian 'troglobites', he announced, was to be found in the physics of evaporation. Cold air can hold very little water vapour compared with warm air. This means that when warm air cools, it rapidly becomes over-saturated and forms clouds of condensed droplets (hence steamy winter breath). Cold air, on the other hand, even if saturated, quickly incurs a saturation deficit as it warms, so that centrally-heated houses feel dry, and cold winds result in chapped lips. In nature, the effect of wind on evaporation is very complex and turbulent airflow across an evaporation surface may produce a disproportionately high water loss. The ambient temperature also influences water loss, rate of evaporation being directly proportional to the saturation vapour pressure at the temperature of the evaporating surface. What this adds up to is that in caves, and particularly in warm, tropical caves, the availablility of water and movement of cave air can result either in drying, or hydrating conditions. The former are more common near entrances, where food tends to be relatively plentiful, while the latter are generally found only in the very deepest parts of the cave, where food may be scarce. Because of their small size and large ratio of surface area to volume, Arthropods are critically vulnerable to desiccation and therefore have been obliged to evolve a suite of physiological, morphological and behavioural characters in order to exploit terrestrial habitats successfully. The most striking requirement of these adaptations is the maintenance of water balance. This can take one of two directions: the elimination of excess water intake, with the necessary conservation of available salt resources; or the development of water conservation mechanisms such as cuticular waterproofing, spiracular control, water vapour pumps, water storage, Malpighian tubules and appropriate behaviour. In other words, arthropods come in one of two separate varieties: either wet- or dry-adapted. What Howarth had found, when he studied his Hawaiian troglobites in detail, was that they were invariably associated with damp, still-air conditions where the relative humidity was typically 100% saturated.

Any animal which remains isothermal with the ambient temperature must continually draw water vapour out of saturated air, since the equilibrium humidity of its body fluids (haemolymph in the case of an insect) approximates 99%. Unless such an animal had effective water excretory mechanisms which conserved salts, it would literally drown in such an environment. Howarth's field observations suggested that terrestrial troglobites in Hawaiian caves are physiologically specialized for life in water-saturated atmospheres, and that this may be the principal feature which sets them apart from their out-of-cave relatives. This observation/hypothesis raised two critical questions: are *all* terrestrial troglobites 'wet-adapted' and if so, *why?* After all, lava cave atmospheres (and indeed most terrestrial habitats in large tropical limestone caves) experience significant evaporation most of the time.

A few years on, we can be fairly confident in answering the first question. Recognizable troglobites *do* seem to be adapted exclusively to humid habitats with water-saturated atmospheres. They generally avoid large, open, draughty passages in macrocaverns, particularly in tropical caves, although other cavernicoles are often abundant in such habitats.

To answer the second question, we need to refer back to the earlier sections in this book on the cave habitat, and recall that caves large enough to admit people form but a tiny proportion of the cave habitat, most of which is of mesocavernous dimensions. Troglobites are specialized for life in the mesocaverns, which are generally damp, frequently flooded and on the whole experience little air-flow.

In 1981, this was brought home to me when, with colleague Mick McHale, I had the chance to join Frank Howarth for a month's work in Kazumura Cave on Kilauea Volcano. As Kazumura was once a major lava artery, we were less than happy when, just half way through our visit, the volcano began its longest and most active eruption this century. Being dedicated scientists (or perhaps out of sheer bravado), we continued with our research, crouching in the angle of the passage when earth tremors threatened to dislodge rocks from the roof. By setting out baits along the cave wall, we succeeded in attracting and studying a wide range of fauna, while monitoring the cave microclimate more or less continuously. As expected, we found that troglobites appeared at the baits when the air was still and moist, but as soon as the cool nightly inward draught began, they quickly vanished into the surrounding mesocavern cracks, where the air presumably remained much more humid and still.

Breezes are a very common feature in caves, particularly in large-diameter tunnels perforated by more than one opening. In general, the larger a cave's diameter and the more extensive its vertical range, the stronger the draught at its entrance. This fact has been appreciated by cave explorers for a long

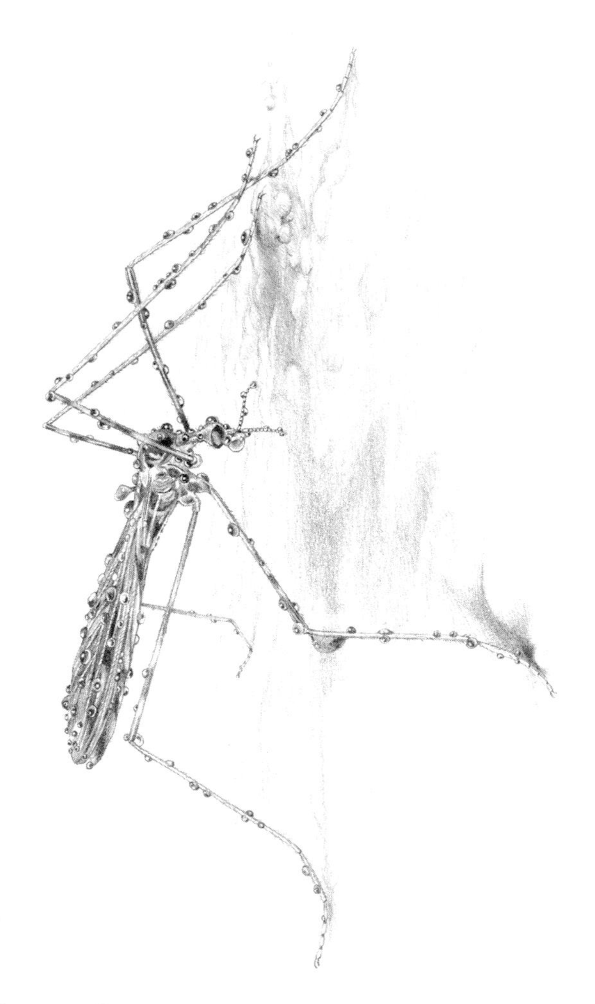

Fig. 2.11 The crane fly *Limonia nubeculosa* successfully aestivates in situations where the cave atmosphere becomes saturated with water vapour (indicated here by a beading of water droplets). At other times, the species shows itself equally tolerant of very dry atmospheric conditions.

time, and many a British system has betrayed its presence by a warm steamy breath puffed from a formerly unregarded crack in the midst of snow-bound moorland in mid-winter.

The cause of such draughts is quite straightforward. Bedrock has a huge thermal capacity, and tends to remain all year-round at the mean annual ambient temperature (making caves an ideal hibernaculum, as we shall see in Chapter 4). Air in a cave within the rock is maintained at that temperature by contact with the cave walls and by convection. In summer, air in the cave is colder, and therefore denser, than air outside. Given the chance, gravity will cause it to flow out of any low entrance to the cave, drawing less dense, warm air into the cave through higher entrances or mesocavernous cracks. In winter, this process is reversed and warm cave air rises out of upper entrances or mesocavern cracks, to be replaced at the lower end of the system by an inrush of cold air. This is known as the 'chimney effect'.

It is winter draughts which most affect cavernicoles. As cold air is drawn into the cave and warmed by contact with the rock walls, it incurs a saturation deficit and begins to dry out the cave. If there is a large vertical distance between entrances, or if the cave has a very large volume, the air-flow may be substantial, leading to severe problems for any cavernicoles who happen to be in the way. Researchers in Mammoth Cave, Kentucky, have reported a winter evaporation rate two hundred times higher than its summer level.

The relative humidity of warm, summer draughts increases as they are cooled by the cave, so that even if the inflowing air starts off relatively dry, it rapidly becomes saturated. In the tropics, especially on mountains (such as the Hawaiian volcanoes), there may be a marked daily fluctuation in above-ground air temperature, from warmer-, to cooler-than-cave. This may result in a twice-daily-reversing draught: upwards at night, downwards in the day. The higher rock temperature in the tropics magnifies the rate of evaporation produced by cool draughts, and the generally large dimensions of tropical cave passages further augments the problem, so that large-volume air flows may traverse several kilometers of cave passages before they reach rock temperature and saturation humidity. For tropical cavernicoles, every night is a mini-winter.

It is not clear to what extent the mesocaverns are affected by draughts, though it would seem likely that in bare karst or lava terrains where they are not soil-capped, some air movement is likely to occur through mesocaverns which connect with otherwise sealed cave systems. Air-filled mesocaverns generally lie at the top of the cave system, with larger-diameter passages at lower levels. Thus, rising winter airflows will usually have reached rock temperature and saturation humidity before they rise into the mesocaverns. Also, as water vapour is lighter than air, it will tend to diffuse rapidly up into mesocavernous cracks overlying moist caves. Summer down-draughts may impact the upper reaches of open mesocavern cracks, but in such confined spaces the cooling air will quickly reach saturation humidity.

Not all parts of all enterable caves are affected to the same extent by climatic variations in the outside world. Howarth distinguishes three biologically-significant climatic zones within air-filled macrocaverns, which he terms '*transition*', '*deep*' and '*stagnant-air*' zones. The first is affected by climatic events on the surface. In the second, the substrate remains moist, the atmosphere remains saturated with water vapour and the cave climate remains stable for extended periods, but air exchange with the surface keeps the air fresh. In the third, air exchange with the surface proceeds more slowly than the build up of gases (especially CO_2) from organic decomposition, resulting in what cavers call 'bad air' (though 'bad air caves' need not necessarily be moist, so Howarth's 'stagnant-air zone' is not a synonym). All these zones may occur in one cave, and the location of the dynamic boundaries between them are determined by such factors as the size, shape and location of the entrances, the size and shape of the cave passages, the availability of water and the climatic regime on the surface.

Climatic zones within the cave may be distinguished by their characteristic faunas. The 'transition zone' is characterized by facultative cavernicoles which have 'terrestrial-type' water regulatory systems, while the more specialised cavernicoles with their 'aquatic-type' water regulatory systems, occur in the other two zones where food resources are adequate to support them. According to Howarth, cave-evolved troglobites are particularly common in the 'stagnant air zone' of caves because this is the macrocavernous habitat whose microclimate most closely resembles that of the mesocaverns to which these troglobites are primarily adapted.

3

Limestone Caves in Britain and Ireland

Karst: the landscape of caves

Caves in Britain and Ireland are nearly all found in one distinct form of limestone terrain, known as 'karst', a Germanic corruption of the Serbo-Croat word 'krš' or 'kras', meaning 'rock'. The word itself probably goes back a long way in the Indo-European family of languages, since it presumably shares the same root as the Welsh or Gaelic 'craig' or 'creag' and the English 'crag'. The original 'Karst' is the Austrian name for a 160 kilometre-wide limestone belt running along the coast of the Adriatic Sea east of Trieste, in Slovenia – an area devoid of surface rivers, a place of barren white limestone surfaces, intricately-fretted rock and yawning cave mouths.

Karst is defined as a landscape which has been developed by water solution of the rocks. More familiar landscapes have been sculpted by surface erosion, and follow an orderly pattern with uplands dissected by a dendritic drainage

Fig. 3.1 Malham Cove in the Yorkshire Pennines was carved by a waterfall flowing from Malham Tarn whose outlet has since adopted an underground course, leaving the cove perched high and dry. (From an engraving by W. Westall, courtesy of Trevor Shaw)

pattern of gradually-deepening valleys which coalesce and widen into lowland plains. Karst scenery, by contrast, seems at first quite disorganized, with hollows and closed depressions of various kinds and streams appearing and disappearing, while quite large valleys may have no rivers at all and may suddenly end without warning in a cliff or a pit. Beneath the surface, the pattern of underground drainage may seem still more chaotic, with cave passages jinking about in three dimensions, diving down below the water table, or narrowing abruptly into impassable crevices. The randomness is more apparent than real. Karst landscapes and cave systems evolve like the living organisms they contain, according to the interaction of natural forces with inherited structures. Making sense of the observed results, or better still, predicting the existence of what cannot yet be seen provides the great excitement of speleology.

Many karst areas are enlivened by considerable, and sometimes spectacular relief, but this is not always so. The Gort Lowlands of the Burren in County Clare lie mostly at or below the 30 m contour line with only very modest relief, yet the region contains many well-developed karst features and caves. Caves can occur right down to sea level (or below in some cases), but most of the great cave areas of the world are in hilly or mountainous terrain. In Britain, limestone caves are virtually confined to hill country to the north and west of a line running from the mouth of the Tees to that of the Exe.

Often the first features to be seen on approaching a karst area are crags – near-vertical walls of limestone with sloping buttresses of scree below them. Most Carboniferous Limestones weather to a light colour, so that such cliffs tend to present an imposing, rather than a forbidding aspect. They are particularly numerous in the Yorkshire Dales where they are called 'scars'. Giggleswick Scar, a faulted escarpment towering over the Skipton to Kendal trunk road, is well known, but there are many other places in Britain with equally impressive escarpments. Two particularly fine examples from Wales are Craig-y-Cilau, overlooking the Usk Valley near Crickhowell, and Eglwyseg Mountain near Llangollen.

Since karst drainage is essentially subterranean, it has been argued that surface river valleys are not karst landforms – yet river valleys are quite often spectacularly developed in karst areas. Those starting on impermeable rocks and then cutting across the limestone are called 'allogenic' or 'through valleys'. Their rivers, compared with those originating on limestone, carry greater quantities of sand or other abrasive material derived from the impermeable rocks upstream and this increases their erosive power; so that through valleys often take the form of deep canyons, or steep-sided gorges.

The development of gorges and scars is helped by the nature and structure of the limestone rock. Most karst-producing limestones are well jointed and often quite thickly bedded. Therefore erosion on slopes tends to be along joint planes with material removed in toppling vertical blocks rather than by the even fragmentation and downhill slumping of surface layers (as happens for instance with shales); and it is for this reason that such magnificent sea cliffs are produced in the Carboniferous Limestones of the Gower and South Pembrokeshire. The V–shaped cross section of a 'normal' river valley results from a balance between the rate of erosion on the valley walls and of removal of accumulated debris by the river at the bottom. The relative rates of these two processes determine how steep-sided the V–section is. Because karstic limestones tend to be hard and massive, and because water drains into the joints of the rock rather than over its surface, valley-wall erosion proceeds relatively slowly in karst terrains. Valley-bottom erosion, by contrast, may proceed quite rapidly, particularly where the river is of allogenic origin, resulting in a gorge.

Limestone gorges can be very impressive. Those of the Tarn and its tributaries in France's Massif Central are on an enormous scale – 1–1.7 km wide and in places 500 m deep. The Vicos Gorge – the 'Grand Canyon of Greece' –

Fig. 3.2 Cheddar Gorge on Mendip is one of Britain's most famous limestone landscapes, now scarred by a busy road. (Courtesy of Trevor Shaw)

runs for 15 km between almost unbroken lines of cliffs nearly 1000 m high. Others, in the Pyrennees, are so deep and narrow as to appear more like cracks in the mountains: the Gorge de Kakouetta is 200 m deep, but a mere three metres wide in places. There are many well-known limestone river gorges in Britain: the Avon at Bristol, Goredale in Yorkshire's Craven district and parts of the Wye Valley immediately come to mind, but the finest in my view is the gorge of Dovedale through the Peak District limestone of Derbyshire. The deeply incised bends and loops in this gorge were formed when the previously mature, meandering River Dove became rejuvenated as a result of uplift of the land, and began actively cutting into its bed once again. Limestone gorges may also be formed by the de-roofing of cave rivers, but this is a relatively rare phenomenon.

When a river crosses from impermeable rocks onto limestone, its waters begin to dissolve down into joints and other fissures so that, over time, a gradually increasing proportion of the flow is swallowed by an expanding underground drainage system. This brings about a change in the pattern of erosion on the surface: downstream of the point where the water is lost, the river bed is worn away less rapidly than it is upstream. Eventually, the whole flow of the river is captured underground, and the now-dry river bed below the sink ceases to be cut down at all, while upstream, the valley continues to deepen. The result is a 'blind valley', closed at its downstream end and forming an enclosed depression.

A typical, well-developed blind valley starts as a normal river valley on the impermeable rocks. It usually continues for some distance on the limestone, perhaps deepening into a gorge, only to end abruptly with its river disappearing into a cave at the foot of a cliff. Good examples are Pant Sychbant in Mid Glamorgan, where the stream disappears into a cave called Ogof Fawr at the base of a wall of limestone, and the Eastwater Valley leading to Eastwater Cavern on Mendip. Indeed many 'sink caves' or 'swallets' (with rivers or streams flowing into them) are situated at or near the ends of blind valleys, but these are often quite small, as in the case of Swildon's Hole on Mendip. The blind valleys of North West Yorkshire are on the whole poorly developed, even though they may lead to large caves. This may result from the very heavy glaciation of the Yorkshire karst until late in the last ice age, so that the present pattern of surface drainage has not been established long enough to have allowed the blind valleys to be deepened to any great extent. Mendip, on the other hand, having been spared the ice-sheets of the last glaciation, has had time to develop deeper blind valleys.

Ireland has some quite good examples of blind valleys. Perhaps the best known is the Coolagh River in County Clare, whose waters collect from the sandstones of Slieve Elva and eventually sink into the flood-prone Coolagh River Cave at Polldonough about three kilometres northwest of Lisdoonvarna. The main part of the Coolagh drainage lies on the shale which overlies the limestone in this area. What seems to have happened is that the river cut its bed down most of the way through the shale until fissures provided it with routes into the underlying limestone, where it swiftly opened up a subterranean drainage route. Collapse of the undermined shales allowed the whole river to pass down into the limestone at Polldonough, and from that time downcutting has been confined to the region upstream, producing a blind valley. This has a V-shaped cross-section, being cut for its entire length into

Fig. 3.3 The Hepste dry river in South Wales. In dry conditions, water sinks into caves along the banks upstream of this point. (Chris Howes)

shales and alluvium, with limestone only exposed at its lowest point near the cave entrance.

At an early stage in the development of a blind valley, the subterranean conduits are only capable of absorbing the river's normal flow, but not the much greater volume which it carries when in flood. At such times the excess water will continue along the old surface bed of the river, and in these cases the valley is termed 'semi-blind'. Such valleys are a common feature of most of our karst areas, and dry river beds may run for miles with no more water than the occasional puddle. It is not uncommon for a river in a semi-blind valley to have not just one, but a whole succession of sinking points for its water, so that the extent of the dry bed varies with conditions of flow. The Manifold in the Peak District illustrates this: for part of the year it fails to maintain a continuous surface flow and under low water conditions may be dry for nearly five miles between Wetton Mill and Ilham Hall. As the flow increases, the river extends beyond Wetton Mill to successive sinks downstream until, all of these having reached capacity, a continuous surface flow is re-established. The effects of a sudden storm can be spectacular in a dry river bed, when a wall of water, perhaps a metre high, instantly transforms the deserted watercourse into a river in spate. I had a narrow escape in Papua New Guinea when such a flood arrived at a 'dry' cave entrance which I had just vacated. The roar of the approaching water provided just enough warning to send me leaping up the bank to safety. Flood pulses of this kind have been reported from the Kingsdale Valley Beck in Yorkshire and the Mellte in South Wales.

Intermittent streams, which carry water only in times of flood, are not confined to semi-blind valley systems. Prolonged, heavy rain may raise the water table high enough to cause springs to flow and streams to run in areas where the normal drainage is entirely underground. This phenomenon is particularly common in the rolling karst landscapes of the chalk. Intermittent chalkland streams are known as 'bournes' and their occurrence may be marked by place names such as Bourne End or Winterbourne.

In most karst areas, including the chalk, there are many valleys devoid of even intermittent streams. Such 'dry valleys' can be recognized by the absence of stream channels in their floors, all drainage having long since moved underground. Accounting for the existence of dry valleys is not always easy. The problem is least when the dry section is clearly a continuation of an existing valley sink. It is more difficult to explain whole ramifying systems which bear no relation at all to the modern karstic drainage, such as are found in the chalklands and parts of the Peak District. Dr G.T. Warwick has explained the Derbyshire case as having been inherited from a pattern of rivers, established when the area was overlain by impermeable shales, which have subsequently become etched into the limestone. Systems of dry valleys might also have originated under periglacial conditions during the Pleistocene, when the land surface was covered by permafrost, as in the present-day Arctic. In the Arctic karsts, the large quantities of melt-water run-off are kept at the surface where they are actively erosive, producing a normal dendritic pattern of fluvial valleys.

There is now fairly general agreement that many of our spectacular dry limestone gorges which were once thought to represent collapsed caves, were cut by melt-waters in times of permafrost. Mendip, which was probably never glaciated but must have been affected by permafrost for long periods, is particularly well endowed, with Ebbor Gorge, Burrington Combe, and finest of them all, Cheddar Gorge. The latter has a well developed system of dry valleys converging on its upper end, including the delightfully-named Velvet Bottom. After Cheddar, perhaps the most imposing dry gorge in Britain is the Winnats near Castleton in Derbyshire, and the nearby Cave Dale is another good example. Further north, in the Craven district of Yorkshire, Trow Gill was probably cut by melt-waters at some stage of the Pleistocene, though part of it may represent a collapsed cavern; however the Yorkshire karst is not otherwise notable for its gorges.

There is no generally accepted English name for valleys which are formed where underground streams break out on the surface, and which are therefore the converse of blind valleys. The Americans call them 'steepheads', which seems an apt term for valleys which are often gorge-like and walled-in at their heads. The best-named British example is the 'World's End' ravine near Llangollen in North Wales. Malham Cove, a huge curved cliff which lies at the head of a small tributary of the River Aire in Yorkshire's Craven district, gives the appearance of a classic 'steephead', but was actually formed by headward recession of a waterfall which once tumbled 100 m from the Watlowes valley above, fed from Malham Tarn. The outlet waters of the Tarn now follow an underground route and the Watlowes is dry, but surprisingly, the water which issues from the foot of the cliff comes from a completely different sink to the west – such is the complexity of karst drainage.

Steepheads are often associated with impressive resurgence caves, such as the show caves of Wookey Hole in Mendip, Peak Cavern in Derbyshire and Dan-yr-Ogof in South Wales. As the valley floor below the cave is cut down, lowering the water table within the limestone, underground streams are driven to find progressively lower outlets so that such caves often have several entrances with the stream emerging from the lowest. 'Resurgence springs' represent the reappearance of allogenic water which entered the limestone at a sink. They are distinguished from 'exsurgence springs' which are fed by rain water which has fallen within the limestone catchment and percolated underground. Most resurgence springs normally also contain some autogenic percolation water, so the distinction is not always clear-cut.

More often than not, springs emerge as upwellings from beneath the water table, or from boulder piles, or from poorly-developed bedding planes rather than from open, enterable caves. Springs which rise from depth under hydrostatic pressure are called 'Vauclusian springs' after the Fontaine de Vaucluse in Provence which forms the source of the river Sorgue. At peak flow in early spring, water gushes up the steeply inclined entrance from the unknown depths of the mountain at the rate of over 500 million litres per hour (140 cumecs). In 1987, a German cave diver, Jochen Hasenmayer, using a special breathing mixture, penetrated 200 m vertically below the surface of the spring. More recently an unmanned probe continued down to a depth of 300 m, at which point the cave seems to level out. Vauclusian springs work like siphons: water is pushed up to the surface by the head of water at the up-flow end of the system, which may be quite distant (the Vaucluse's sinks are 40 km from the rising). Vauclusian springs are not confined to mountainous regions, but may even appear on the sea bed, as they do in the Bristol Channel close to the islands of Steepholm and Flatholm, or in the intertidal zone, as at Kinvarra in Western Ireland, where at least part of the Gort River resurges. Some Vauclusian springs are warm, presumably because the water rises from considerable depths, and some have such a high dissolved mineral content that, as the emerging water cools or loses excess CO_2 to the atmosphere, excess lime is precipitated as tufa. Knaresborough in Yorkshire and Matlock Bath in Derbyshire are noted for 'petrifying wells' where tufa is deposited on objects, such as bird's nests and top hats, which visitors have tied on to the rock, so that they are suspended in the warm spring water.

I have already referred to the apparently chaotic pattern of relief in karst landscapes, and perhaps the major feature contributing to this chaos is the pock-marking of their surface by closed circular or oval hollows called 'dolines'. Stretches of the South Wales karst are so heavily pitted that from the air they look as though they have been subjected to blanket-bombing of a ferocious intensity, and most other karst areas in Britain have similar, if less spectacular markings. In section, most dolines are bowl-shaped or conical, but some are relatively deep and narrow and are then commonly called 'potholes' or 'pots'. Potholes usually have bare rock walls, while most dolines are soil-lined and vegetated, often with loose boulders at the bottom.

Dolines form by a combination of solution and collapse and are all related to underlying cave (or mesocavern) systems which provide their drainage – although only a few contain open entrances large enough to admit a caver. Dolines in Britain range in size from a few, to a hundred or so metres across, and may be tens of metres deep, but elsewhere in the world they may reach massive proportions. The forest-filled Luse doline in New Britain is 750 m wide by 230 m deep, with a volume of 60,000,000 m^3. The Crveno Jezero, a part-flooded pot in southern former Yugoslavia, is 400 m in diameter, ringed by 160 metre-high cliffs and, contains a lake 270 m deep. But perhaps the most impressive dolines of all are the Blue Holes of the Bahamas, one of which, the kilometre-wide Black Hole of South Andros Island, contains an unknown depth of cold sea water plunging into blackness below a warm surface layer of fresh water.

Where the limestone is thickly covered with permeable alluvium or glacial drift, as in North West Yorkshire, the gradual subsidence of the underlying rock produces shallow dolines known locally as 'shakeholes'. These are easily distinguished from the steep-sided 'collapse dolines' formed when surface layers of rock tumble down into well-developed cavities and which are most common where the limestone is overlain by another impermeable rock, as in parts of the north crop of the South Wales coalfield.

Most dolines drain only their immediate neighborhood, but some capture surface streams and become active sink-holes, known on Mendip as 'swallets'. However not all dolines are free-draining: inwashed clay may seal the outlets allowing a lake to form, or the water table may rise so that the doline becomes a large natural well, or 'cenote'. The water in cenotes is usually more or less static, but some dolines have been formed by collapse of the ceilings of river caves, where the resulting pothole has water rushing across its floor. Such dolines tend to elongate along the course of the river and are known as 'karst windows' – a famous example being the 300 metre-deep entrance into the Nare River Cave in New Britain which drops the unfortunate caver into the middle of a raging rapid, with a wet-season flow of around 700 million litres per hour (200 cumecs).

Fig. 3.4 Dolines above Little Neath River Cave, South Wales. (Chris Howes)

Largest of all the closed depressions in karst are 'poljes' – large flat-floored valleys which typically have springs around their margins, giving rise to streams which cross them before vanishing once more into sinkholes or caves. They take their name from the Serbo-Croat for a cultivated field, for in the mountainous Dinaric Karst, the rich alluvium of the poljes provide the only land fit for cultivation. In wet weather the sinks may be inadequate to drain the polje which then floods. Popove Polje, just inland from Dubrovnik, is typical in this respect. In winter it is a huge lake; in summer a mass of intensively cultivated fields, with the villages perched on the surrounding rocky hillsides – seeming to spurn the convenient flat land below, but in fact merely being prepared for the winter.

The British karst has little to offer in the way of poljes except perhaps for some shallow hollows in the Hale Moss area of Morecambe Bay. In the Burren of Western Ireland, however, the Castletown River flows through swampy ground across part of the 5 km^2 Carran depression, before sinking at its southern end. A few kilometres west of Carran are other features which might also be classed as poljes, including the intermittently-flooding Kilcorney depression which contains the romantically named but very muddy Cave of the Wild Horses. Similar karst depressions in the Gort Lowlands of the Burren fill or empty with bewildering speed, reflecting rapid fluctuations in the water table. These are the legendary 'turloughs',from the Irish 'tuar lough' meaning dry lake. All British turloughs are found in Ireland, except for one recently discovered in West Wales. Some of the turloughs nearest to Galway Bay also fluctuate with a tidal rhythm. The most remarkable of these is Caherglassaun Lough which, under some conditions, has a tidal range of several metres, even though it is 5 km from the sea and has no surface stream flowing either into or out of it. The perplexing behaviour of the lake probably has something to do with the final underground course of the Cannahowna, or Gort River, a most confusing waterway which sinks and rises repeatedly and has a variety of names and several changes of direction between Lough Cutra and Coole Lough. Evidence of a subterranean connection with Caherglassaun Lough is provided by the fact that when the Gort River is in flood, the Lough is not only full, but loses its tidal oscillations; when the river is low, on the other hand, the Lough drains completely at each low tide.

Apart from its loughs, the Burren is famous for its superb 'limestone pavements' – extensive, near-horizontal slabs of naked limestone scoured clean by the ice sheets of the Pleistocene. Similar pavements occur in the Yorkshire Dales, where solutional etching and carving have opened up the joint cracks in the limestone to form drainage fissures known as 'grikes'. Where two sets of joints intersect more or less at right angles, as is often the case in the Yorkshire karst, the blocks in between – known as 'clints' – are substantially rectangular. Yorkshire clints average perhaps two metres by one, with a lot of local variation, while grikes range in width from a few centimetres to a metre or so, and in depth from 50 cm to three metres or more. Grikes often accumulate soil and may shelter a wealth of specialized plants.

As solution proceeds, it not only widens and deepens the grikes, but often undercuts the clints by opening minor bedding planes resulting in a surface of loose slabs which may rock when stepped on. Solution may also attack minor fractures in between the joint sets, giving the edges of the clints a delightful crinkled appearance – just one example of the many types of minor

Fig. 3.5 The spectacular limestone pavements of the Burren, Co. Clare. (Chris Howes)

solutional patterns collectively known by the German word 'karren'. In the tropics, karren can assume some weird and wonderful forms such as the soaring razor-edged pinnacles of Gunung Api in Sarawak, or the vertical boxwork of jagged rock blades characteristic of Madagascar's 'tsingy' karst. But karst erosion surely reaches its most majestic heights in the clusters of 'fenglin' towers which soar up out of the Li Jiang river plain in southern China's Guangxi Province. While we cannot boast anything on quite the same scale in our islands, we do have a few impressive karst pillars, such as Pickering Tor, Tissington Spires and Ilam Rock in Dovedale, the pinnacles in Cheddar Gorge and the Seven Sisters Rocks and Devil's Pulpit in the Wye Valley.

The emotional response to karst scenery must be a personal one – to me karst landscapes are not just exceptionally beautiful, they also hold an element of excitement at the thought of the secret wonders which lie below the ground, just waiting to be discovered. It is surely no accident that nearly all the true karst of England and Wales lies within the boundaries of either National Parks or designated Areas of Outstanding Natural Beauty.

Caves in Britain and Ireland

To describe, or even to merely list the 3000 or so documented caves which occur in the Carboniferous Limestones of Britain and Ireland is clearly beyond the scope of this book. Yet our subject demands that I should attempt to convey something of the nature, variety and 'feel' of the places where cavers disport themselves and cavernicoles live. Caves conceal some of the least accessible habitats in the British Isles, and any naturalist who wishes to study them must first become a competent caver. An intimate relationship between exploration, sport and science has always been a feature of speleology. Most major cave systems have been discovered bit by bit over a long period of time through the cordinated efforts of succeeding generations of dedicated individuals, who are prepared to spend endless hours digging through glutinous mud in cold, cramped, wet and often dangerous conditions. In opening up caves to human exploration, cavers have changed them profoundly by destroying speleothems, removing sediment blockages, modifying patterns of subterranean drainage, airflow and microclimates. So the story of their exploration is inextricably bound up with the character of the caves themselves.

In the following pages, I shall attempt to convey something of the general character of the caves found in our major cave regions, illustrated in each case with reference to one or a few typical or notable cave systems described from the perspective of the explorer-speleologist.

Mendip

Though by no means the most important caving area in the British Isles, the Mendip Hills have been a focus of subterranean exploration since people first began to take an interest in caves. There is plenty of archaeological evidence to show that it was prehistoric man who made the first tentative exploration of the caves of Mendip during the later stages of the last Ice Age. The dry caves

Fig. 3.6 The downstream side of Rowten Sump in Kingsdale Master Cave (Yorkshire Pennines) epitomizes the caver's dream - a mysterious, living cave with immense potential for discovery. (Judith Calford, photographed by Chris Howes)

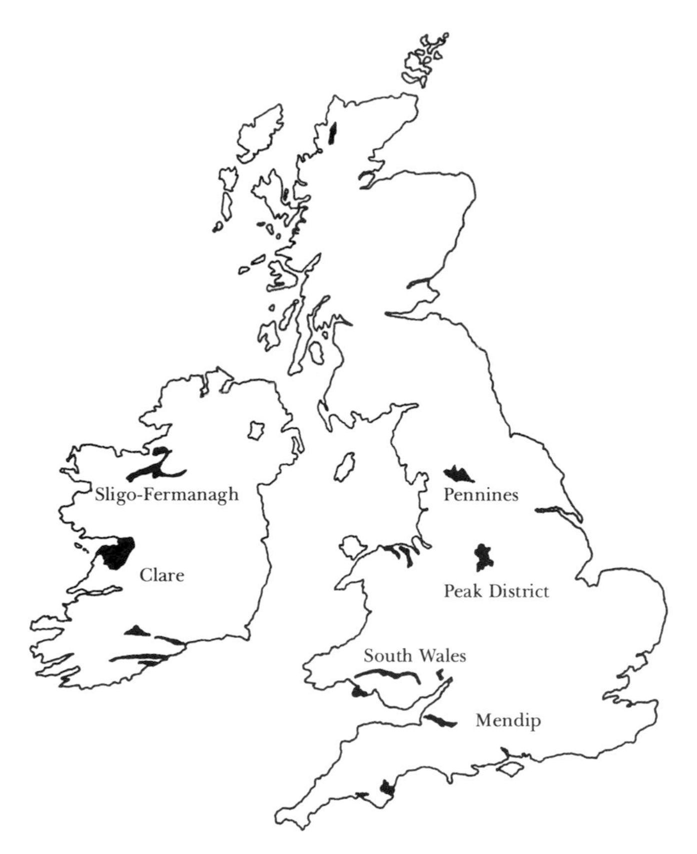

Fig. 3.7 Main cave areas in Britain and Ireland (after G.T. Warwick in *British Caving* ed. Cullingford, Routledge and Kegan Paul).

of Burrington, Cheddar and Wookey provided many convenient refuges both for domestic purposes and for toolmaking. Popular tradition refers to these people as 'cave men', but it is more likely that they were nomads and only occasionally inhabited caves. In Romano-British times the caves were again in use, both for domestic and business purposes – there is evidence that one cave was used for manufacturing counterfeit coins. The early visitations were probably not true cave explorations, in the sense of being motivated by curiosity about the caves themselves – except perhaps in the case of Wookey Hole, where the fourth chamber was used as a burial site in Roman times. Travelling a distance of nearly 200 m, the first visitor to this chamber must surely have been a caver at heart.

The documented record of exploration in the Mendip caves begins with the activities of lead miners, who accidentally encountered natural cavities while tunnelling for profit – the discovery of Lamb Leer in the 17th century being a prime example. Scientific study and excavation of cave sites by Boyd Dawkins and other gentleman archaeologists of the 19th century gave cave exploration an air of respectability and provided the inspiration to H.E. Balch (1869–1958), one of the most famous of the caving pioneers.

The formation of local caving clubs at a very early date gave impetus to both the sporting and the scientific sides of caving, with the result that Mendip is perhaps the most intensely studied karst region in the world – although the formidable amount of effort invested has yielded pitifully few major cave discoveries. Mud blockings are all too common and the dedicated Mendip caver will spend much of his life digging at surface or underground sites. These digs may last for decades and digging is thirsty work, as noted in the lyric of at least one cavers' song (to the tune of 'The Wild Rover'): "There's no caves on Mendip – or so runs the tale, so potholing prowess is measured in ale!"

Not quite all the caves of Mendip were massively blocked at the entrance when first investigated. Swildon's Hole, a stream sink near Priddy, had a bouldery obstruction just a few metres underground, but this was passed with a small amount of effort by R. Troup and his companions in 1901 to enter what is today Mendip's finest stream cave. Troup and party cautiously descended a rift and soon found themselves following the cave stream through a small hole into a water chute, where they received a thorough soaking. The streamway, and the cavers, now plunged downwards through a delightful series of potholes and small waterfalls until a deep pool, with flowstone crowding down from above, forced a regrouping. In 1901, the water reached almost to the roof, but Harold Hiley plunged in and dug out an accumulation of debris from the far side, providing a much more reasonable airspace. The baggage was passed through and the party continued their descent. Shortly after, the stream entered a roomy chamber with a bouldery floor – the Water Chamber. A wide inlet passage entered from the right, while ahead the stream disappeared into a narrow, muddy rift – the Water Rift – at which the party halted.

The following weekend, the group was joined by Balch, who nicely expresses the difficulty posed by the Water Rift to these pioneer cavers, burdened as they were with bulky equipment and a hand-held candle for lighting:

"When we first arrived at this spot we were confronted by deep water and deeper mud beneath. To attempt to wade was to be bogged, and it was seen to end not many yards ahead, with stalactite filling the passage, its pendants being submerged so that it looked impassable. The rift narrowed upwards and there were inconspicuous ledges on the stalactited walls. Getting up between these, with feet on one side and shoulders on the other, the stalactite barrier was reached, and it was discovered that there existed, high above the water, a tiny passage through it. With the maximum of effort, if one had not dined too well, and nature had not been too generous with one's circumference, head and shoulders were thrust in and keeping the body up in the sloping and narrowing creep, one presently emerged over the continuation of the water."

Within a few metres a waterfall pitch was reached – the Forty Foot Pot – which, compounded by the hostile attitude of the farmer, put an end to further progress for the next ten years. In 1914 the farm was sold. The new owner made the cavers welcome and that summer a party including E.A. Baker laddered the Forty and followed a large stream passage to the head of another pitch – the Twenty.

At this point war intervened, and no significant exploration was made until the drought year of 1921. A new attitude towards exploration was soon apparent in a series of 'improvements' to the cave: the 'Mud Hole' of the Water Rift was filled in with rocks and the bottom of the stalactite barrier broken away, making it much easier to get tackle to the Forty. It is hard for modern cavers

Fig. 3.8 Early exploration in Swildon's Hole, Mendip. H.E. Balch is seated, second from the left. (Photo by Harry Savory, courtesy of University of Bristol Spelaeological Society).

to imagine the sheer volume of bulky baggage these early explorers took with them, but they were commonly underground for twenty hours, and on occasion far longer. To quote Balch again:

"To negotiate creeps, the counted packs are passed from hand to hand to be dumped near the first man of the party, which then moves forward till the last man reaches the dump, when the operation is repeated, with a count on every occasion to prevent loss."

In 1921, the Twenty was descended and the streamway followed under stalactite barriers and along tall rift passages to the Double Pots, two water-filled potholes of larger than usual dimensions. Baker, who "often climbed where a cat would fall", traversed their edges, but the rest of his party were stopped. The Double Pots were regarded as a major obstacle and future expeditions, usually led by Balch, sent an "amphibious man" ahead to rig a handline. The streamway held no more major obstacles until a sump was reached, a pool floored with decaying leaves and mud from which methane gas bubbled when they attempted to dig it out.

Another decade passed with little advance except in equipment, whereby Balch's "dependable illuminant", the candle, was gradually abandoned in favour of carbide lights which, though they cast "treacherous shadows", provided far more light. Electric lights were also used but did not become truly popular until reliable and waterproof ex-Coal Board lights became available in the 1960s.

An account of the siege upon Sump 1 would take several pages, if justice were to be done to the story; here a short summary will have to suffice. The main protagonists were Graham Balcombe, Jack Sheppard and C.W. (Digger) Harris. Suitable diving equipment was unobtainable, so the first attempts on the sump were made with explosives. In 1932, Harris placed a small charge detonated by a battery and alarm clock after the party had left the cave. The results were minimal, and a later second charge, for which a shothole had been laboriously drilled, was equally unsuccessful. An attempt was then made to dive the sump, with air drawn through a length of garden hose to a mouthpiece made from part of a bicycle frame. Balcombe located the underground passage, but had difficulty drawing air against the water pressure. Sheppard then tried, but the mouthpiece came apart and he narrowly escaped drowning. The explorers reverted to explosives, using 10 lbs of gelignite held under the arch of the sump by a frame of bamboo canes. The explosion, at one o'clock in the morning, woke people in the village of Priddy and loosened an area of the roof above the sump pool. A week later, two more charges were detonated, the second of which disturbed evensong in the village church. The sump, however, remained, and the use of explosives was abandoned.

A second attempt at diving the sump involved pressurized air, fed from a small hand pump. Equipped with rubberized garments, Sheppard passed the sump in 1936. It proved to be a mere two metres long, and a fortnight later Balcombe was the first to free-dive it (without breathing equipment). He and Sheppard explored the open streamway beyond two low ducks (cave passages with limited airspace) to Sump 2. Four weeks later, wearing a home-made diving suit fed from a cylinder of compressed oxygen, Balcombe passed the 10 m long Sump 2 to reach an air bell beyond which lay Sump 3. This sump, being deep and somewhat constricted, proved impassable to the bulkily-clad divers of the day. It was to remain so for another twenty years.

The breakthrough came, as so often in caving, in a totally unexpected place. A beautiful grotto in the roof of the cave, upstream of Sump 1, had been found by E.K. Tratman during an early Balch expedition. Although a passage could be seen continuing beyond it, Balch had pleaded for its preservation and it became known as Balch's Forbidden Grotto. In 1945, a group of cavers smashed a way through the grotto, but were stopped by a stalagmite boss. This was blasted away in 1953 and an extensive complex of high-level passages was eventually explored. In a distant corner, the sound of running water provided the motivation for a further two years of digging and blasting down the narrow 'Blue Pencil Passage', which gave access, as anticipated, to the main streamway of what is now Swildon's Four. Here the water burbled over gravel flats and in and out of rocky pools, through the finely sculptured rift of the 'Tate Gallery'. The inevitable Sump 4 was easily passed by divers in 1958, but has since become choked with gravel and is now awkwardly tight.

Sump 5 was passed in 1958 and three years later, Mike Boon, taking advantage of the more compact technology of the aqualung, managed to get through the constricted Sumps 6 and 7, before turning his attentions to the equally tight Sump 3, which he passed without difficulty. The route down the streamway was now open and the long and arduous porterage via Blue Pencil Passage was consigned to history. Sumps 8, 9 and 10 fell in 1965 to a dynamic group of young divers: Mike Wooding, Dave Drew and Dave Savage. The trio then found a bypass to Sump 11, but were halted by the deep and constricted Sump 12, which remains unpassed to this day.

In 1968, a dramatic flood cleaned out a gravel-choked passage in Swildon's One, providing an easy bypass to the Forty Foot Pot. Equally dramatic changes in equipment and attitudes since that time now mean that a trip to Sump 1, considered such a serious undertaking in the 1920s and '30s, is now regarded as a suitable introduction to caving for novices. Beyond Sump 12, the Swildon's water is lost – at least temporarily. Two kilometres to the south, it reappears in Wookey Hole as the underground River Axe.

The exploration upstream from Wookey Hole is a story as long as that of Swildon's, but while the latter is a cave mainly above water, the former is mainly underwater. A dry route to Chamber 3 has always been open and forms the original show cave with its Witch of Wookey stalagmite. Chambers 4 and 5 are only accessible during low water when a dam (at the entrance to the cave) is released.

In 1935, Graham Balcombe and Penelope Powell, equipped with lead boots, metal helmets and long tubes for delivering air, made the first underwater expedition to reach Chambers 6 and 7. The war put a halt to further exploration, but during the war years an oxygen rebreather diving system was produced for the navy. Rebreathers differ from the now more familiar SCUBA equipment in that the diver breathes a gas mix through a closed loop, where exhaled breath is chemically purged of carbon dioxide, topped up with oxygen and then fed back to the diver to be breathed again. After the war, surplus rebreather kits were snapped up by the newly formed Cave Diving Group. By 1948, a large dry chamber had been found above Chamber 9. The deep sump beyond seemed impassable, for the pure oxygen supplied by early rebreather systems becomes toxic at depths below nine metres. An answer seemed to lie in the compressed-air 'aqualung-with-fins' system favoured by

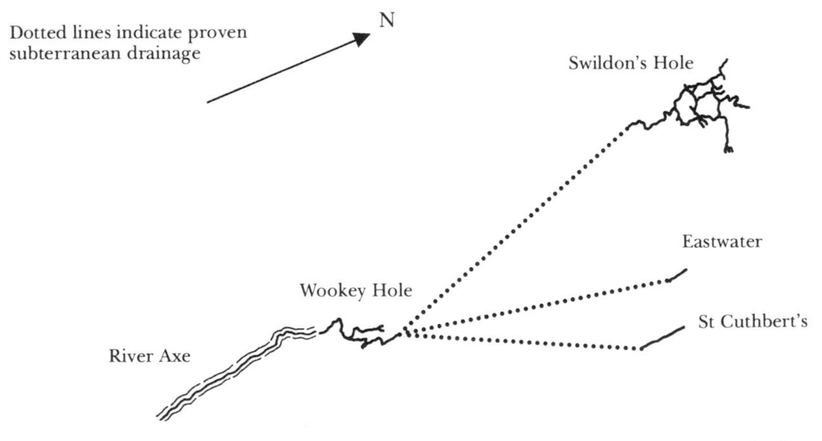

Fig. 3.9 The relationship of Wookey Hole to nearby caves whose waters feed into it.

Fig. 3.10 One of the earliest photographs of cave divers in action using back-mounted cylinders to explore the flooded section of Wookey Hole, on Mendip. (Luke Devonish)

French divers and this was tried by Bob Davies in 1955 – with near-disastrous consequences. He lost his guide-line in a fog of sediment and ended up in a new, unknown airspace. With great presence of mind, and considerable luck, he waited for four hours until the water had cleared, then found his own way out. Aqualungs gained an unjustly bad reputation as a result of this incident and for a while further progress was left to the bottom-walking rebreather divers, using nitrogen-oxygen mixtures rather than pure oxygen, to increase their potential depth.

Mike Boon's success far upstream in Swildon's in 1962 sparked a revolution in cave diving and within two years the aqualung was universally adopted, together with the neoprene wetsuit – a tight-fitting garment of bubble-filled rubber with good insulating properties which trapped a layer of water next to the skin for added warmth. The days of woollen-clothed, goon-suited rebreather divers were ended.

The 21 metre-deep sump of Wookey 9–18 (with its nine associated above- and below-water chambers) was passed in 1966 by Dave Savage, and in 1970 John Parker found the large dry cave of Wookey 20 after a dive of 150 m from the last air surface in Chamber 9. The nomenclature of the Wookey chambers is horribly confusing, with numbers being given to each new discovery whether above or below water. The dry passages of Chamber 20 proved to lead nowhere, so attention turned to the Chamber 20 sump pool and a possible inlet passage underwater. A further dive of 150 m upstream from Chamber 20 led Parker to airspace in Chamber 22. A mined tunnel, extending the show cave to Chamber 9 now provided a new advance base for diving operations. Improvements in both equipment and techniques allowed longer and safer dives to be made; divers now carried two or more lights and at least two air cylinders, each with independent valves and mouthpieces.

In 1976, Colin Edmunds made the long-anticipated breakthrough from Chamber 22 into an underwater inlet passage heading upstream towards Swildon's. Two days later, northerners Oliver Statham and Geoff Yeadon, to the disgust of local divers, pushed on through a series of short and shallow dives to the magnificent Chamber 24, where the whole of the River Axe came pouring towards them along a passage 12 m high. Six hundred metres later, they reached a deep blue lake from which the water welled. A few days later Edmunds and Martyn Farr returned, passing the deep sump to reach the desolate confines of Chamber 25 – 'The Lake of Gloom'. An awkward rib of rock divided the lake in two. On its upstream side lay the deepest and most serious sump of all.

The sump beyond Chamber 25 is so remote and so deep that new techniques have had to be developed, more akin to those of deep-sea diving. Farr and, more recently, Rob Parker have made heroic attempts to pass it, using a range of breathing mixtures of oxygen, nitrogen and helium in order to minimise the time required for decompression in the numbingly cold water. In 1985, Parker reached a British record depth of 67 m, where he was stopped by the full force of the River Axe gushing up through an impassably low slot. The quest for a subterranean link with Swildon's had for the time being come to an end.

Pennines

The Pennines, centred around the four peaks of Gragareth, Whernside, Ingleborough and Pen-y-Ghent is Britain's major caving area. Its 1800 or so caves and potholes require three large guidebooks to expound their virtues, and it is estimated that on a normal Saturday more than 500 cavers can be found in pursuit of its speleological delights.

Developed in flat-lying, massive limestone beds, the caves typically have a stepped profile where long horizontal passages alternate with plunging vertical shafts. The classic Pen-y-Ghent Pot provides one of the finest examples. Although only 194 m deep, including the depth of a dive in the terminal sump, getting to the bottom of the cave involves long crawls and twelve vertical drops or pitches, accompanied all the way by an icy stream which floods the system after heavy rain.

Perhaps the most impressive pothole in the Dales is Alum Pot, on the eastern slopes of Simon Fell. The deep green entrance shaft is wide enough for daylight to penetrate right to the floor, over 70 m below. A small stream cascades down one wall, filling the lower regions with a misty spray which adds to the feeling of awe experienced by the caver as he or she dangles over the lip on a suddenly all-too-slender rope.

Gaping Gill on Ingleborough is probably Britain's best-known pothole. The powerful stream of Fell Beck is swallowed by a dark gash in its bed, giving no indication of the huge drop below. One hundred and ten metres down, it crashes onto the shingled floor of the largest cave chamber in the country, from which around 15 km of passages link with ancient tunnels at a number of levels to form a system whose complex history has not yet been fully unravelled.

The most extensive cave network in Britain, known as the Three Counties System, lies north of Ingleton in a horseshoe curve around the southern end of Gragareth. Its main components are the Ease Gill-Leck Fell complex beneath Casterton Fell and Leck Fell, straddling the boundary between Cum-

Fig. 3.11 A caver abseils from daylight into Juniper Gulf – a classic pothole system which drops 130 metres below the moorland of Ingleborough (Yorkshire Pennines) in three vertical steps. (Chris Howes)

bria and Lancashire; the Notts Pot-Ireby complex underlying Ireby Fell in Lancashire; and the West and East Kingsdale complexes in Yorkshire. Between them, these caves contain around 110 km of known passages. A highly simplified, schematic map of the system is given in Fig. 3.12, from which it can be seen that quite a few of the supposed links between main sections of the system are based on dye traces through flooded tunnels and have yet to be physically explored by cavers. Nevertheless, proved connections in the western part of the system have brought the explored length of the Ease Gill-Leck Fell network to 70 km, making it Britain's longest cave, and at 211 m deep, the deepest in the Pennines.

In broad terms, the components of the system grade from the inclined, dendritic vadose networks of Ease Gill, to the vertical-vadose-plus-horizontal-phreatic systems of Ireby Fell and West Kingsdale. All the caves are contained in the near-horizontal beds of the Great Scar Limestone, which form a broken plateau looking out towards Lancaster and Morecambe Bay. The Dent and North Craven Boundary Faults delimit the limestone to the west and south, while to the north the outcrops are capped by shale. During the Pleistocene the whole area was overrun from the north by ice sheets and two major glaciers

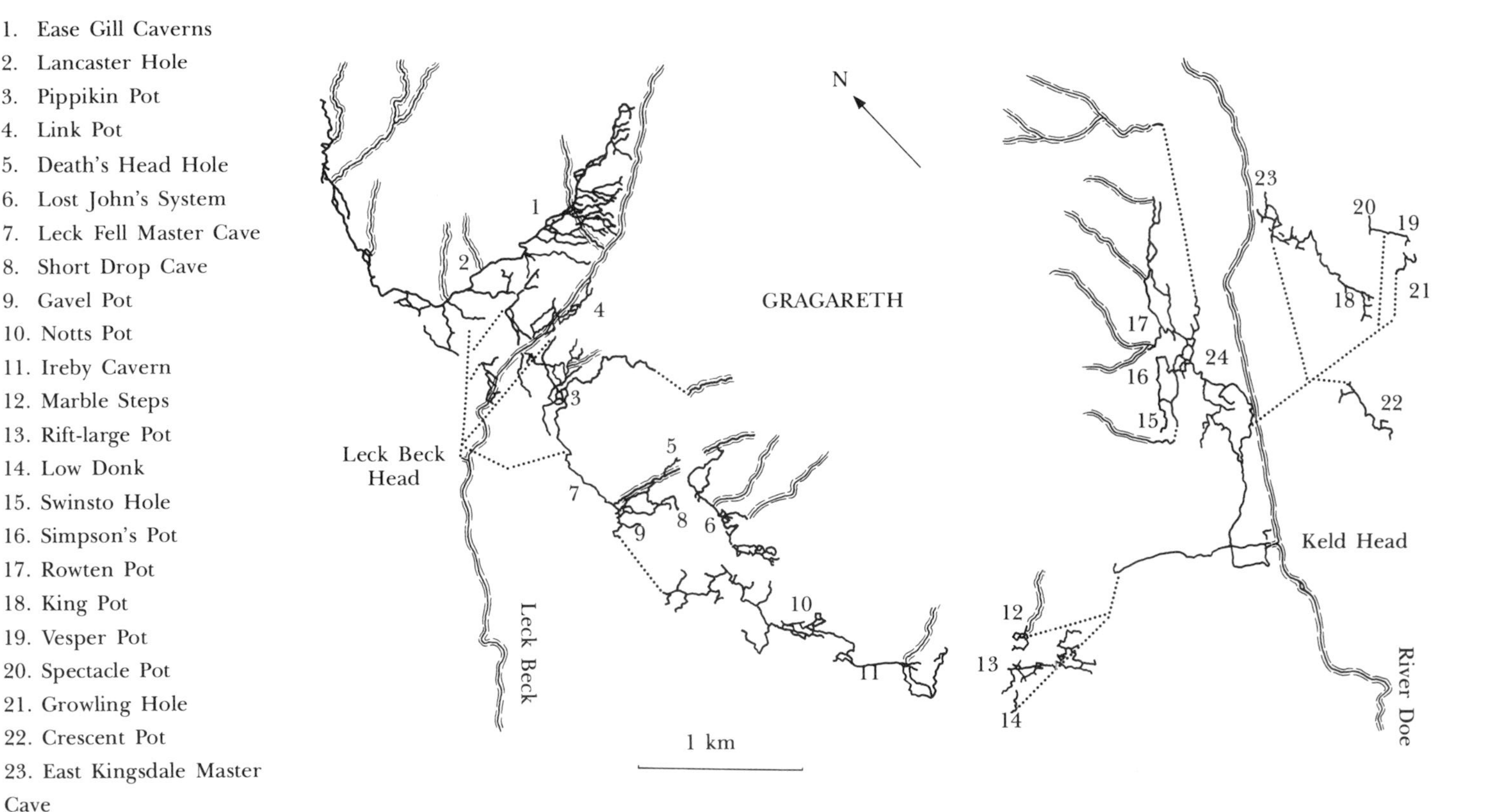

Fig. 3.12 The Three Counties System, Pennines (after an original survey compiled by D.B. Brook).

flowed either side of Gragareth, gouging the U–shaped troughs of Kingsdale to the east and Barbondale to the west. Overflow from the Barbondale glacier carved out Leck Beck, partly infilling the Ease Gill valley with fine sediment as the ice retreated.

The karst drainage of the area is dominated by streams running off the shale cap and sinking into the limestone. Once underground, the cave waters make their way to one of two resurgences – Keld Head in Kingsdale, or Leck Beck Head at the foot of the Ease Gill Valley. Interconnection of the caves within the catchment of each resurgence has long been considered as a possibility by local cave divers – and over the last twenty years, the links have gradually been made one by one, often in epic circumstances.

The discovery of the underwater route between Keld Head and the Kingsdale Master Cave by divers Geoff Yeadon and Oliver Statham stands as one of the classic feats of modern exploration. Following the pioneering efforts of Mike Wooding in 1970, in 1975 Yeadon and Statham began a series of long and difficult dives in Keld Head, swimming against the current in water at a chilly 8°C, in conditions of near-zero visibility, at a depth of around 10 m. By the spring of 1976, they had laid guidelines in the flooded passages beyond the resurgence to a horizontal distance of 700 m. As this was about as far as they could go given the capacity of their diving tanks, they now switched to the Kingsdale Master Cave, hoping for a downstream route to link up with the end of their guidelines. On July 1, 1976, Yeadon reached 400 m downstream in difficult passage, and was on his way back, when his demand valve suddenly flooded.

"Having just breathed out, I quickly reached for the purge button but found none. All that was left was the rubber mouthpiece, the main body having dropped off. With lungs now demanding air in increasingly urgent twitches, I started to feel for my second [spare] valve, as the visibility was too poor to see it. The first attempt followed up the wrong hose to the contents gauge. The second was successful."

Undeterred, he continued to extend the route past some very tight underwater squeezes, to 640 m.

On August 5, 1976 Oliver Statham claimed the British cave diving record when he reached a point 838 m into Keld Head. Yeadon claimed it back with a dive of 920 m at greater depth and in appalling visibility, and so it continued.

By 1978, equipped with larger cylinders giving a greater range, Yeadon had advanced the downstream limit in Kingsdale Master Cave to 731 m and had descended a dark and gloomy pot to a depth of 18 m. The upstream and downstream limits were now at roughly the same depth and the divers' survey suggested that a connection was a mere 60 m away. On July 6 of that year, Statham and Yeadon were joined by the German diver Jochen Hasenmayer for a final 'big push'. The divers took turns to extend the line, and in an epic dive lasting 2 hours and 30 minutes, Yeadon at last made the historic connection. On January 16, 1979, with television cameras in attendance, the pair made the 1830 m dive from the cave to the resurgence, emerging from Keld Head to a champagne reception.

While the endless watery maze beyond Keld Head and Leck Beck Head may hold the key to the eventual integration of the Three Counties System, the impression which most cavers take home from a visit to the area has little to do with silent, flooded tunnels; for this is pothole country.

In his book *Underground Britain*, Bruce Bedford gives a vivid description of Lancaster Hole – discovered on a calm day in 1946, when a trembling tuft of grass led members of the British Speleological Association to their greatest discovery:

"The main trunk passage, huge beyond belief in parts, lies beneath Cumbria's Casterton Fell, and contains many indications of the last ice age. The legacy of those violent times is evident in colossal hillocks of fractured rock, in rolling waves of compacted mud, and a jumbled wedge of rubble 30 m above the present-day stream, which burbles beneath glowering caverns in a splendid cockled waterway, oblivious to the devastation that forms its roof."

The adjacent Ease Gill Caverns, first entered in 1947, consists of a number of dendritic sink caves fed by leakage from the Ease Gill stream, all flowing in a roughly north-to-north-west direction to join the main streamway which flows westwards towards Lancaster Hole. For cavers who enjoy the challenge of route-finding, the opportunities for adventure in this system seem endless. Jagged rents in the roof lead to old wide caverns festooned with stalactites and vadose canyons twist and turn, seeming to continue for ever until progress is halted by a vertical tube, or aven, soaring upwards into darkness, or an impassable slot. Low, wet crawls lead to distant chambers far upstream, hours away from the well-worn routes. For those who do not know the system well, it is possible to spend hours trying to find the way on, or out, and rescues are not infrequent.

Pippikin Pot lies on Leck Fell, just across the Ease Gill valley from Lancaster Hole. It was opened by blasting in the 1960s and is much loved among cavers for the twisting, body-sized tubes and rifts of its entrance series, which pose as much a mental as a physical challenge. Having successfully negotiated a sequence of squeezes and pots, the caver enters an extensive labyrinth a hundred metres or so beneath the moor. In 1979, the Northern Pennine Club dug into the convoluted maze of Link Pot, between Pippikin and the Ease Gill System, and soon had joined them together, making it possible for 'dry' cavers to pass from Leck Fell in Lancashire, beneath the Ease Gill Valley, to emerge on to Casterton Fell in Cumbria.

South of Pippikin lies Death's Head Hole, a fine 65 m shaft leading to a segment of fossil trunk cave which drains into the lower end of the Lost John's System. The entrance series of Lost John's contains a choice of fine vadose passages alternating with clean-washed shafts which open into the Leck Fell Master Cave. Downstream this enters the phreas, where it is joined by the water from Short Drop Cave, via the rejuvenated phreatic tunnels of Gavel Pot. The upstream sump in Gavel has been dived to a rattling shingle blockage at a depth of 64 m, dye traced to have come from the caves beneath Ireby Fell.

Across the moor to the south, Notts Pot is a vertical maze of vadose shafts which lead to a short stream passage. Downstream this has been dived to Notts II – with 4 km of passages, parts of which come within 100 m of Lost John's. Upstream, divers have made the connection to Ireby Cavern, a cave of long graded streamways with a few short waterfalls at its entrance. Swinging east, the dramatic sinkhole of Marble Steps straddles the present underground watershed between the catchments of the Kingsdale and Leck Beck valleys. To its south, the Rift-Large Pot system consists essentially of a massive, deserted tunnel which is the eastwards continuation of the Duke Street tunnel at the

Fig. 3.13 A forest of straw stalactites decorates the ceiling of Easter Grotto in Ease Gill Caverns. (Chris Howes)

bottom of Ireby Fell Cavern. This tunnel probably carried the waters from the lake which occupied part of Kingsdale during one or more interglacial periods, westwards towards Leck Beck. The immature, flood-prone rifts and shafts of Marble Steps and Low Douk now drain east to the Keld Head rising.

The West Kingsdale System is the most completely explored segment of the Three Counties System, and represents a classic example of active cave development. Three streams, at Rowten Pot, Simpson's Pot and Swinsto Hole, offer exceptionally pleasant caving down a series of vadose shafts and canyons to their junction in the West Kingsdale Master Cave, and then underwater to Keld Head. A fossil passage heading east from the Master Cave has been cut through by the glacial valley, providing a dry low-level exit to the system.

For the confirmed masochist, King Pot on the opposite side of the valley provides a long, cold, wet trip, with a lot of crawling and ten vertical pitches leading into the East Kingsdale Master Cave. Higher up the same hillside, Vesper Pot, Spectacle Pot and Growling Hole are all small streamways which choke in fault-guided rifts. Their streams continue through to link with those from Crescent Pot and the Master Cave and the combined waters then flow completely underneath the valley in a phreatic tunnel, joining the main drain of the West Kingsdale system on its way to the Keld Head rising. A team of divers including Geoff Yeadon and Geoff Crossley have succeeded in linking the East and West Kingsdale systems in a record-making 3.5 kilometre-long dive lasting over six hours.

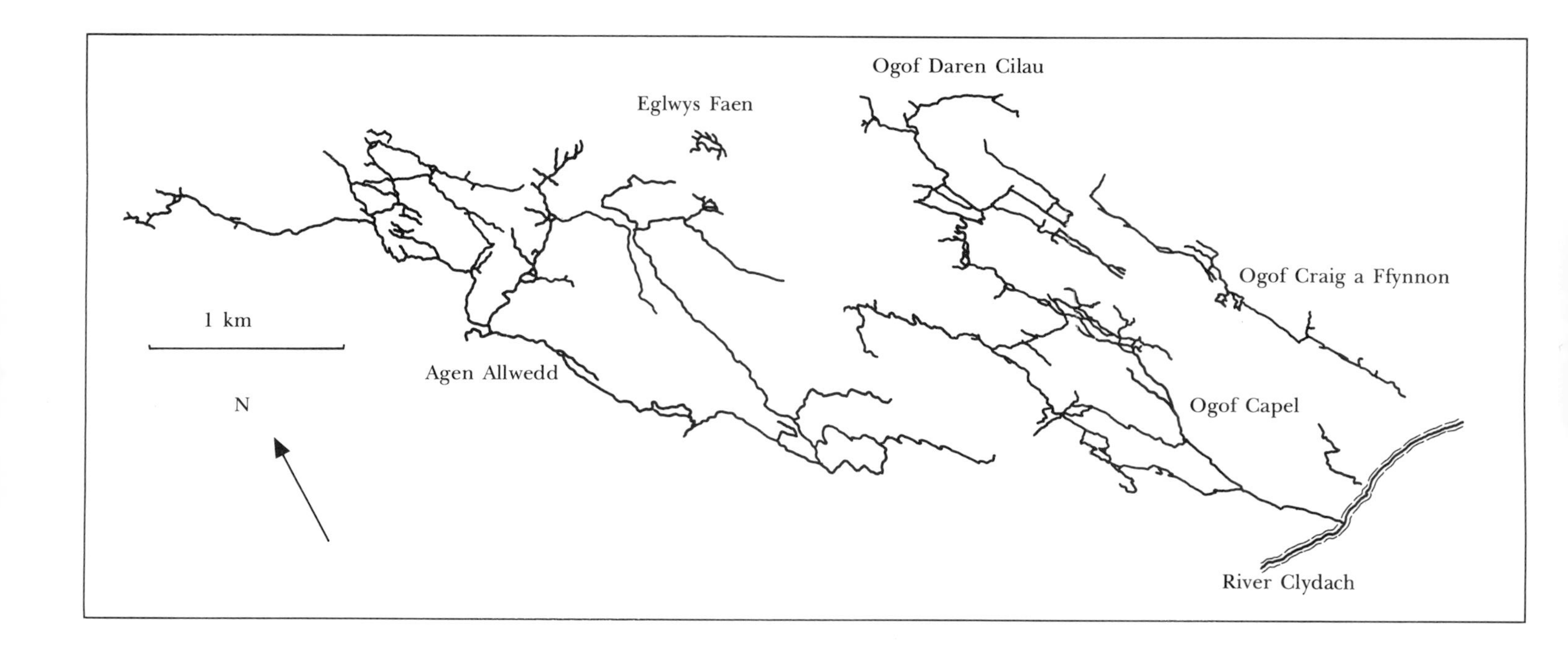

Fig. 3.14 The Caves of Mynydd Llangattwg (after P.L. Smart and C.G. Gardener in T.D. Ford *Limestones and Caves of Wales.*, C.U.P.)

South Wales

The Carboniferous Limestones around the northern rim of the South Wales coalfield are not very extensive, but on Mynydd Llangattwg and in the Tawe (Swansea) Valley, they contain some of Britain's longest and finest caves.

The peat-covered Langattwg escarpment lies at the eastern end of the Brecon Beacons, separated from the Black Mountains to the north-east by the valley of the River Usk. The limestones here are overlain by Millstone Grit, which, with the Coal Measures, caps the whole of the 'Valleys' coal-mining area to the south. On Llangattwg, the whole unit dips gently southwest at an angle of 2–3°, and the springs which drain the limestone, instead of being sensibly placed in the valley below, are tucked right around the southern corner of the scarp in the gorge of the Clydach River. As a result, the underground drainage runs away from the Usk Valley and into the escarpment, picking up most of its recharge waters by vertical percolation through joints in the overlying Millstone Grit.

The Mynydd Llangattwg System currently comprises three major caves: Ogof Agen Allwedd (32 km long), Ogof Craig-a-Ffynnon (8 km) and Ogof Daren Cilau (25 km). For a long time, Agen Allwedd, better known to cavers as 'Aggy', was the only cave on Llangattwg worth visiting. Entered from a small, excavated entrance at the base of the crag, the miserably low-roofed entrance series leads into the huge 1200 metre-long Main Passage. This is a segment of a complex of bone-dry fossil passages, now partly choked by breakdown and sandy sediments. A short way along Main Passage, the Main Stream Passage leads off for a kilometre or so in a south-westerly direction to a junction where it is joined by the Turkey Streamway, flowing south. Upstream, this leads to the large sandy tunnels of Summertime. Below North West Junction, the Main Stream Passage offers a kilometre of wet caving through two major boulder chokes to reach the Lower Main Streamway which ends in a long sump. A return can be made via the unremittingly dreadful Southern Stream Passage, which offers two kilometres of continuous crawling, stooping, crab-like squeezing and wriggling past obstacles, most of the time in cold water.

Ogof Craig-a-Ffynnon, named after the nearby 'Rock and Fountain' pub, consists essentially of two enormous isolated segments of a collapsed fossil trunk passage, running above a series of exceptionally linear passages heading north-west towards the entrance of Ogof Daren Cilau. The magnificent Hall of the Mountain Kings was first reached in 1977 by John Parker, Jeff Hill and Anne Franklin, after a determined four year onslaught on the entrance boulder chokes using gelignite, crowbars and sweat. The following year saw the breakthrough into the even bigger Way of the Kings (Ffordd-y-Brenhinoedd), festooned with eccentrically-twisted stalactites, known as helictites, and crystals, and the kilometre-long Promised Land passage, with its curious pagoda-like calcite formation. Over the last decade, the pace of discovery has ground to a virtual halt, with efforts now centred on the fifth boulder choke.

Following the opening of its tight and awkward entrance series in 1963, Ogof Daren Cilau remained for many years the *bête noire* of Llangattwg, generally disdained and infrequently visited. However, on August Bank Holiday, 1984, Clive Gardener and two colleagues from the Chelsea Speleological Society, lured by a mysterious draught, dug through boulders in the wall of

Fig. 3.15 The Time Machine passage in Ogof Daren Cilau is the largest natural underground passage in Britain. (Clive Gardener)

the old Main Rift and entered a tight lower passage with an even more promising breeze. The breakthrough came a fortnight later, with the party strengthened by the addition of Martyn Farr and Paul Tarrant. A boulder blockage was removed and the explorers were rewarded with three kilometres of large, open passage, including the magnificent Epocalypse Way, of similar dimensions to the 'Kings' passages in Ogof Craig-a-Ffynnon, complete with its own calcite 'pagoda'. The anticipated connection failed to materialize, but on a subsequent visit in March 1985, Tony White completed a bold climb in another part of the new extension and discovered the south-trending White Passage which was followed up a boulder slope to the start of the Time Machine. The first glimpse suggested a vast chamber with shadowy voids off to distant walls – in fact that 'chamber' proved to be 400 m long and by far the largest underground passage in Britain.

Over the following year, major discoveries followed one after the other in rapid succession. A bypass was found to the terminal choke of the Time Machine, leading to a small streamway decorated with intricate helictite 'Bonsai trees' and long calcite straws. Ahead, the stream was swelled by the 'lost' waters of Epocalypse Way. Beyond Bonsai Streamway, the Kings Road led on southwards to a major stream junction, which sumped both ways. The upstream St David's Sump, which is preceded by a roomy underground lake, was passed by Martyn Farr in April 1985, and pushed for 800 m to a further sump, the Gloom Room, with hopes for a future connection to the final sump of Agen Allwedd.

From the start of the Kings Road, some hard digging in late 1985 and early 1986 rewarded the self-styled "Hard Rock Crew" with the next major advance – the Rock Steady Cruise – a dry and occasionally choked walking-sized passage which led into what became known as the Hard Rock Extensions. Beyond the St David's Sump, divers entered the Borrowed Boots streamway where 1500 m of unrelenting hard caving led eventually to the Seventh Hour Sump, described by its discoverers, Rob Parker and Ian Rolland, as "undoubtedly the most remote sump in the British Isles". This was subsequently by-passed via a climb to reach the Inca Trail – heading straight towards Trident Passage in Agen Allwedd. To date the connection remains elusive but no doubt one day, with the aid of diving and perhaps a modicum of explosive, the Llangattwg cavers will forge a single super-system to rival the Three Counties System of the Northern Pennines.

For connoisseurs of underground waterways, the Tawe Valley is nothing short of Mecca. On the west side of the valley, at the foot of Castell-y-geifr, lies Wales' only show cave, the beautifully decorated Dan-yr-Ogof where a stream emerges from a gaping cave mouth. Beyond lies a series of lakes and underground waterfalls, and 15 km of passages, some exquisitely decorated with every type of calcite formation.

Across on the east side of the valley, tucked away among some trees, a rusty steel door marks the insignificant lower entrance to Ogof Ffynnon Ddu – 'the

Fig. 3.16 The Top Waterfall in Ogof Ffynnon Ddu's streamway. (Chris Howes)

Cave of the Black Spring' – known to the faithful simply as OFD. Two additional entrances up on Carreg-lwyd between them give access to five kilometres of splendid, crashing streamway which drains a network of high-level passages extending for upwards of 35 km. A vertical range of 306 m makes this Britain's deepest cave.

OFD is one of the most complex caves in Britain, both physically and in terms of its geomorphological history. The structure of the cave has been strongly influenced by fractures in the rock which are aligned north-south, and by the dip of the limestone beds, which varies considerably in direction and steepness, but trends roughly south at 6–16°.

From a caver's point of view, the streamway is a delight. Wide enough to maintain the water level between knee- and waist-depth, the passage snakes away into the hill, its ceiling soaring to unseen heights. The walls, cockled by the chemical bite of the torrent, glisten black, and are streaked here and there with the lightning-like zigzags of pure white calcite veins. In times of flood, the stream roars and foams in echoing mayhem, banking the curves like a bobsleigh and sweeping all before it. This is a time for sensible cavers to seek shelter in the network of higher passages, for in such a mood the cave has taken lives.

The early explorations of the cave in the 1940s were unusual in that they began at the bottom of the system, at an excavated entrance near the the Black Spring itself, and worked their way upwards. At first the streamway yielded less than 400 m before a boulder collapse and a series of sumps halted progress. Dip Sump was passed by divers Charles George and Brian de Graaf in 1960, but six more years elapsed before the way on was found from Shower Aven into OFD II, with its two kilometres of magnificent streamway and its network of high-level passages. The sump has now been by-passed and a trip through this part of the system reveals a feast of speleothems – arrays of stalactites and stalagmites, delicately-twirling helictites and crystal pools, and – tucked away in a distant *cul-de-sac*, mirrored by a shallow pool – the Columns, perhaps Britain's most famous underground formations. Over a dozen slender pillars of white calcite span the two metre height of the passage, set off by the curve of the wall and embellished with slender straws and stubby, red-tipped stalagmites; they cannot but inspire wonder in the beholder.

With the finest streamway in the country and formations of exquisite beauty, Ogof Ffynnon Ddu was declared a National Nature Reserve in 1976 – the first cave in Britain to merit such recognition.

Derbyshire's Peak District

The bleached limestones of the White Peak occupy about one third of Derbyshire's Peak District National Park. To the north they are bounded by the sombre, peat-covered grits, shales and mudstones of the Dark Peak, culminating in the wild moors of Kinder Scout. The limestone measures some 40 km by 15 km and rises to 450 m at Eldon Hill, its rolling Dales concealing a multitude of caves and mines.

During the Lower Carboniferous, the Peak District site was covered by a large tropical lagoon, fringed by coral reefs and surrounded by deeper water. Massive lagoonal limestones form most of the White Peak, with more irregularly bedded reef limestones cropping out along the edges, as at Treak Cliff, near Castleton.

At the end of the Lower Carboniferous, crustal movements raised the limestones clear of the sea, while simultaneous movements far to the north bucked huge mountains skywards, spilling great rivers down over a floodplain whose sediments would harden as the shales and grits of Edale and Kinder Scout.

The major caves of Derbyshire are clustered along the contact between the impervious shales and the fissured limestones in the saddle above Winnat's Pass, west of Castleton. They include such popular holes as Perryfoot and Giant's, Gautries and Dr Jackson's caves and, closer to Castleton village, the well known tourist show caves of Peak Cavern and Speedwell Mine. Though several of these caves begin, or end, in the irregularly bedded reef limestones (Peak Cavern's huge entrance 'The Vestibule' being a classic example), most of the major cave development has occurred in the more massive lagoonal limestone which forms the bulk of the hills.

A stroll eastwards along the B6061 from Sparrowpit will take the cave-spotter past a line of stream sinks, the first of which enters Perryfoot Swallet. The eighth, appropriately named P8, or Jackpot, is a popular novices' cave. Its fine vadose streamway is punctuated by a six metre pitch which becomes a crashing waterfall after rain. Lower down, the cave opens out before entering a series of sumps, nine of which have so far been passed by divers, yielding a total passage length of over 700 m. The stream disappears into the sixth of these sumps, to reappear at Main Rising or Whirlpool Rising in Speedwell Cavern from which it proceeds to its resurgence at Russet Well in Castleton Village.

Further along the sequence of sinks, P12 enters the well-known Giant's Hole. The first organized explorations here were mounted by the British Spelaeological Association in the mid 1950s. After a prolonged siege, BSA members passed a series of difficult crawls and wet sections near the entrance, to reveal over three kilometres of dry caverns and active streamways. Modern explorers have an easier time of it since the landowner blew open the restricted crawls in a frustrated attempt to fashion a tourist show cave. A four and a half metre pitch leads to the 'Crabwalk' – perhaps the most infamous cave passage in the Peak District. Seven hundred and fifty metres long, this sinuous trench is so narrow at its base that normal forward motion is impossible. Instead the caver must adopt a sideways wriggle-and-shuffle technique. Further along the cave, a temporary slackening in gradient has allowed the stream to open out the base of the rift to the more comfortable dimensions of the 'Great Relief Passage'. More crawls and drops eventually lead to East Canal, and another partially explored sump, some 160 m below the entrance.

In the high levels of Giant's Hole, an unpleasant succession of belly crawls are relieved by the unusual calcite formations of 'Poached Egg Passage', which leads to more awkward squeezes and ducks in the 'Chamber of Horrors'. In 1966 this section was connected to the Oxlow Caverns, a part-natural, part-excavated system entered via an old mineshaft in the hillside opposite Giant's Hole. The entrance pitches drop down in large mined-out natural cavities. The fourth pitch lands in 'West Antechamber', from which a short crawl leads to the huge 'West Chamber', which is also intersected by the Maskhill Mine via 152 m of unstable pits and excavations. The combined Oxlow / Giants / Maskill System is the longest and deepest in Derbyshire, with 4800 m of passages spanning a vertical range of 242 m, making it the second deepest system in the British Isles.

The most impressive pothole in the area is Eldon Hole, situated some two kilometres from Giant's, on the slopes of Eldon Hill – the highest summit in the White Peak. Its impressive maw, 34 m by 6 m in area and 85 m deep, is formed on a master joint within a block of reef limestone. Not surprisingly, it has inspired many local legends. A goose flying into Peak Cavern (known as the "Devil's Arse in the Peak") was said to have emerged from Eldon Hole with blackened wings, having flown too near the fires of Hades. The descent of Eldon Hole by J. Lloyd and E. King, presented to the Royal Society in 1772, was the first objective account of a cave exploration in Britain to be accompanied by a reasonably accurate survey. Despite the fears of the local villagers who had lowered him to the bottom on ropes, Lloyd emerged from the cave "unscathed and not blackened by the fires of hell".

An earlier descent of the cave by an expendable peasant at the bequest of the Earl of Dudley is described by Thomas Hobbes in *De Mirabilis Pecci* (1683).

"Tis said great Dudly to this Cave came down,
In fam'd Eliza's Reign a peer well known.
He a poor peasant, for a pretty price
With rope around his middle does entice
And pole in hand, like to Sarissa tight,
And basket full of Stones down to be let
And pendulous to hang i'th'midst o'th'Cave
Thence casting stones intelligence to have
By listning, of the depth of this vast hole.
The trembling wretch descending with his pole
Puts by the stones, that else might on him rowl
By their rebounds casts up a space immense
Where every stroak does death to him dispense
Fearing the thread on which his life depends
Chance might cut off, ere fate should give commands
After two hundred yards he had below
I'th earth been drowned, far as the Rope would go,
And long enough hung by't within the Cave,
To th' Earl (who now impatient was to have
His answer) He's drawn up, but whether fear
Immoderate distracted him, or 'twere
From the swift motion as the Rope might wreath,
Or Spectrums from his fear, or Hell beneath
Frighted the wretch, or the Soul's cittadel
Were storm'd or taken by some Imp of Hell,
For certain 'twas he rav'd; this his wild eyes
His paleness, Trembling, all things verifies.
Where venting something none could understand
Enthusiastick hints ne're to be scan'd,
He ceasing dies after eight daies were gone.
But th' Earl inform'd how far the Cave went down."

(At this point a footnote explains "To wit to Hell"!)

Peak Cavern must rival Wookey Hole as Britain's best known cave. During the 17th century and beyond, it was used as a factory-warehouse by a colony of ropemakers who lived in rows of cottages inside the huge 30 metre-wide, vaulted entrance, connected to the medieval Peveril Castle on the hillside above. Beyond the entrance, the cave extends for over 13 km and includes some of the finest examples of phreatic passages in Britain, some with an almost perfectly symmetrical cross-section.

The cave achieved modern notoriety as the scene of one of the most ill-publicized cave rescues of all time. In 1959, Neil Moss, an Oxford undergraduate, was trying to pass an unexplored vertical slot in a passage beyond Pickering's Passage when he became hopelessly jammed. All attempts to free him failed and he eventually suffocated when carbon dioxide built up in the pit below. His body was never recovered and the remains were concreted into their natural tomb.

The water flowing through Peak Cavern under normal conditions is fed by autogenic percolation from the moors above, but the cave can also act as an overflow for flood waters from the P1–P4 cave streams and Speedwell Cavern. When this happens, the normally dry entrance series in Peak can fill up very rapidly and there have been instances when the mud and sand dumped by the receding floodwaters have completely sealed off access to the cave. The sumps from which these floodwaters rise have provided a happy hunting ground for cave divers since the pioneering explorations of Donald Coase and Graham

Fig. 3.17 The classic phreatic tube of Peak Cavern's main passage. (Paul Deakin)

Fig. 3.18 The canal level in Speedwell Cavern was constructed to bring out lead ore from the mined part of the system. (Courtesy of Chris Howes)

Balcombe in the mid 1940s. Diving connections have been made with nearby Speedwell Mine and other long-abandoned mines whose surface entrances had collapsed or been lost. These 'discoveries' underline the extent to which "t'owd man" (as the old miners were popularly known) had explored subterranean Derbyshire long before cavers arrived on the scene. The water which enters the depths of Speedwell Cavern from caves such as P8, behaves in a most peculiar manner, alternating between two risings a long way apart in the system. For a time, all the water comes from Main Rising, then there is a flood, and afterwards it all comes from Whirlpool Rising. This goes on for a few months or even years, until another big flood magically switches the entire flow to Main Rising again.

Peak Cavern is owned by the Duchy of Lancaster and is run as a show cave from March to October each year, with concrete paths and artificial lights to ease the way. Visitors to the nearby Speedwell Cavern are even more pampered by the owners, being conveyed by boat along a subterranean canal originally designed to float ore barges out of the mine. The canal eventually reaches a natural cavern called the 'Bottomless Pit' which was reputed to have swallowed 4000 tons of mine waste without altering the water level. Beyond the Pit, cavers can explore over three kilometres of river passages and dry upper levels originally formed by the waters from Giant's Hole and P8, on their way to Peak Cavern.

The history of metal mining in Derbyshire goes back to pre-Roman times, and, as on Mendip, is closely bound up with the exploration of caves. Most of the commercially valuable deposits of galena and zinc blende infilled pre-existing limestone caves and mesocavernous cracks during the Trias – the period, incidentally, when some of our present-day troglobites are thought to have first occupied caves. In his search for mineral wealth, "t'owd man" was, in many cases, re-excavating some of the oldest caves in the British landscape! The Good Luck Lead Mine at Matlock is a working museum which celebrates the culture of Derbyshire mining, and the tourist section of Bagshawe Cavern, at Bradwell, southeast of Castleton, also includes some mined sections.

Apart from its heavy metallic ores, the Peak is also famous for a strikingly beautiful form of calcium fluoride, or fluorspar, known locally as 'Blue John'. It occurs as thick veins infilling the spaces between large blocks of a fossil collapsed cave system in the hillside just north of Winnats Pass. The workings at the Blue John Cavern and nearby Treak Cliff Cavern are open to the public.

Sligo-Fermanagh

A glance at the geological map of Ireland should gladden the heart of any karst enthusiast, for it shows a central basin of Carboniferous Limestone underlying almost half the country. The best caves are concentrated in the uplands of the Burren in County Clare and in those parts of Counties Sligo, Leitrim, Cavan and Fermanagh flanking the Arigna Coalfields.

In County Sligo the caves have formed in the youngest part of the strongly chertified Upper Limestone known locally as the Dartry Limestone. Concealed in the wild expanses of open bog are many rifts and potholes, including the 142 metre-deep Carrowmore Pot in the Geevagh area, until recently the deepest known cave in Ireland.

The more extensive caves of County Fermanagh, are developed in the clean, massive 'reef' facies of the Dartry Limestone, whose structure encourages the development of pothole systems similar in profile to those of the Yorkshire Dales. Despite their remote setting, the Fermanagh caves have a long history of visitation. De Latocnaye in his book *A Walk Through Ireland* (1796) described in colourful detail his cave explorations made while guesting with the First Earl of Enniskillen. In 1895 Edouard Martel included Fermanagh in his grand tour of Britain and Ireland and his lectures later inspired the Yorkshire Ramblers to carry out much of their early pioneering work in the county.

The caves are situated in two main areas southwest of the county town of Enniskillen, along the northern edges of Tullybrack and Cuilcagh mountains, partially split by the Lough MacNean valley. Tullybrack is noted for its classic potholes leading into great horizontal systems while Cuilcagh is particularly famed for Marble Arch Cave – a splendid dendridtic system which drains into the head of the picturesque Cladagh Glen Nature Reserve.

Reyfad Pot, the deepest system in Ireland with a vertical range of 179 m and the third longest with 6.7 km of passages, lies high on the northeast slopes of Tullybrack in a tree-lined depression on the open bogland. It stands as one of the great unknowns in Irish caving with only a fraction of its potential realised. Great gaps exist on the survey between known cave passages and its dye-traced resurgence at Carrickbeg. The pothole's larger entrance swallows a small stream draining off the shales above. Inside, an eight metre pit, formed along a fault, leads through a small window into the impressive 60 m main shaft. Having negotiated a boulder constriction at the bottom of the spray-lashed pitch, the potholer emerges in the roof of a magnificent passage with a 24 m abseil to the floor. This, the main trunk route, is also aligned along a fault and measures over 15 m by 15 m. Much of the rest of the cave consists of passages of similar dimensions, heavily infilled with glacial deposits of mud and boulders which frequently completely block the passage and thwart further exploration. Progress is made by clambering up and down over mountainous sand banks or squeezing along the gap between infill and roof created by shrinkage and compaction of sediment. In other places streams are slowly carving out the deposits to reveal the original passage in all its glory. It is obvious that this is an extremely old system dating back to pre- or early-glacial times with the modern streams utilising the ready made passages.

In the fraction of passage so far discovered there are at least three streams pursuing independent courses. One sinks at Polltullybrack, a young postglacial cave which presents the opportunity for a classic 'exchange' trip. One hundred and fifty metres of flat-out crawling in water through a passage studded with sharp-edged nodules of chert rock, leads into the roof of the Reyfad Main Aven, where a 54 m abseil sets up a sporting trip onwards to the bottom of the shafts below the main entrance. The second stream sinks into an inconspicuous and at present inaccessible pothole at Pollnacrom, while the third and largest ('New River') drops directly from the surface into the part of the system known as the the Shower Room and disappears almost immediately into an impenetrable boulder choke. The trend of all streams is southerly, while the resurgence lies two kilometres to the east, offering an exploration challenge for the future.

The cave system draining the adjacent watershed is the notorious Noon's Hole, named for Dominic Noon, an informer who in 1826 was cast into its supposedly bottomless depths as punishment for his treachery. Nowadays it boasts the more favourable reputation as one of the finest and most sporting through trips in the British Isles.

The entrance is a small sink at the corner of a field, its insignificance belying the hidden grandeur beneath. It was Martel who, in 1895, first dammed the stream and made a partial descent to 35 m, reporting that below him the pot belled out into a "colossal giant's cauldron" of fluted, water-washed walls. In the early 1900s, the Yorkshire Ramblers passed Martel's limit to reach a series of further ladder pitches, broken by bridges and ledges, which gave an intricate route down to the floor of a gloomy chamber, 92 m below the entrance. It was a remarkable exploration feat for the time, but the celebrations were short-lived, as the way on from the bottom was halted almost immediately by a sump. The Yorkshiremen had overlooked the key to the system which lay at the base of a small parallel fossil shaft leading off from the Main Ledge. It was left to the Leeds University Speleological Society to make the breakthrough to the stream cave beyond in 1970. The vadose passages quickly sumped downstream below a series of debris-strewn pools, but upstream the cave yielded extensive passages, including the inevitable High Noon and After Noon Series.

In 1973, divers Martyn Farr and Roger Solari turned their attention to the resurgence, the splendid Arch Cave concealed in a densely wooded valley one and a half kilometres from the sink as the crow flies. They passed two sumps linked by smooth-walled canals to arrive in a huge "starless void". It was a cave diver's dream – one and a half kilometres of stunning river passage which they named Arch 2. They dived the eventual upstream sump to arrive at the bot-

Fig. 3.19 Abseiling into the entrance of Pollasumera, a short stream cave which feeds the 6.5 km-long Marble Arch cave system in Co. Fermanagh. (Tim Fogg)

tom of Noon's Hole with the teasing glimmer of daylight filtering down from the unreachable entrance far above. In 1984 a dry connection into the Arch 2 Streamway was found via a circuitous route from High Noon, allowing non-divers to appreciate what is indeed a classic trip.

Twenty kilometres to the south, the 667 metre-high Cuilcagh Mountain dominates the landscape, its cliff-lined sandstone ridge rearing from stepped terraces, the lowest and widest of which is formed of the Dartry Limestone. At its western end the source of the River Shannon has been traced to the 2.1 km long Shannon Cave, while its eastern end conceals a wealth of deep pots. In between lies the classic Marble Arch System which first attracted the attention of Martel and subsequently the Yorkshire Ramblers. Carved in the 'reef' limestone, it boasts six and a half kilometres of sweeping river passages and chambers of lofty dimensions. There are three main sinks on the open bog just beyond the Namurian/Limestone contact: Pollasumera, a beautiful yawning entrance but with little open passage beyond; Monastir, where water is engulfed at the base of a 40 m high amphitheatre and thirdly the Sruh Croppa. Access to the system is gained by a number of wet or dry entrances just above the rising where the three streams emerge as the River Cladagh at the contact with the underlying thinly bedded Glencar Limestone to flow under the magnificent cave remnant the 'Marble Arch' itself.

With a folding canvas boat Martel explored from the 'Wet Entrance', rowing upstream to an area he called the Junction and Grand Gallery where the three streams finally combine. His exploration was halted by a sump, a bypass to which was discovered in 1935 by the Yorkshire Ramblers and dubbed the Flyover. It led into the Pollasumera River passage which they named Skreen Hill 1 – a beautifully decorated section ending in a lake and sump. It was not until 1965 that Mike Boon of the Shepton Mallet Caving Club moved a few boulders in the wall of the final lake to reveal a squeeze up into a curtain-draped chamber and a fossil bypass to rejoin the river at Skreen Hill 2. This phase of discovery also disclosed a further one and a half kilometres of tributary passage, Legnabrocky Way, which includes Giant's Hall – one of the biggest chambers in Ireland measuring 60 m by 30 m by 15 m. The Marble Arch 'jigsaw' was completed in 1967 when the Irish Caving Club dived through to Skreen Hill 3, discovering 650 m of splendid stream passage still accessible only by diving.

In 1985 the Wet Entrance and Skreen Hill 1 were adapted as a magnificent tourist cave, fulfilling a similar suggestion made by Martel in his book *Irlande et Cavernes Anglaises,* published nearly a hundred years earlier.

Clare

The main caving area of County Clare lies to the north-west of the county in a region known as the Burren – from the Irish 'boireann' meaning 'a rocky place'. It is aptly named, for during glacial times, ice sheets ground across the northern hills, bulldozing their soils down onto the southern lowlands to produce extensive moraines, and leaving behind a landscape stripped to its bare rock bones. The vast, clean-scoured limestone pavements of the Burren constitute one of the finest glaciokarst landscapes in the whole of Europe and harbour the richest limestone flora in the British Isles.

The geology of the area is simple. The Burren limestones dip gently to the south, passing beneath impermeable shales and sandstones. Three outlying shale-capped summits provide the catchments for the majority of the known caves. Of these summits, Slieve Elva and Knockaunsmountain are connected to the main body of impermeable rocks by a thin bridge of shales. In the vicinity of Lisdoonvarna, the shale bridge is less than 20 m thick, and some surface streams have sliced through it to reach the underlying limestone. Immature caves are presently forming in their beds.

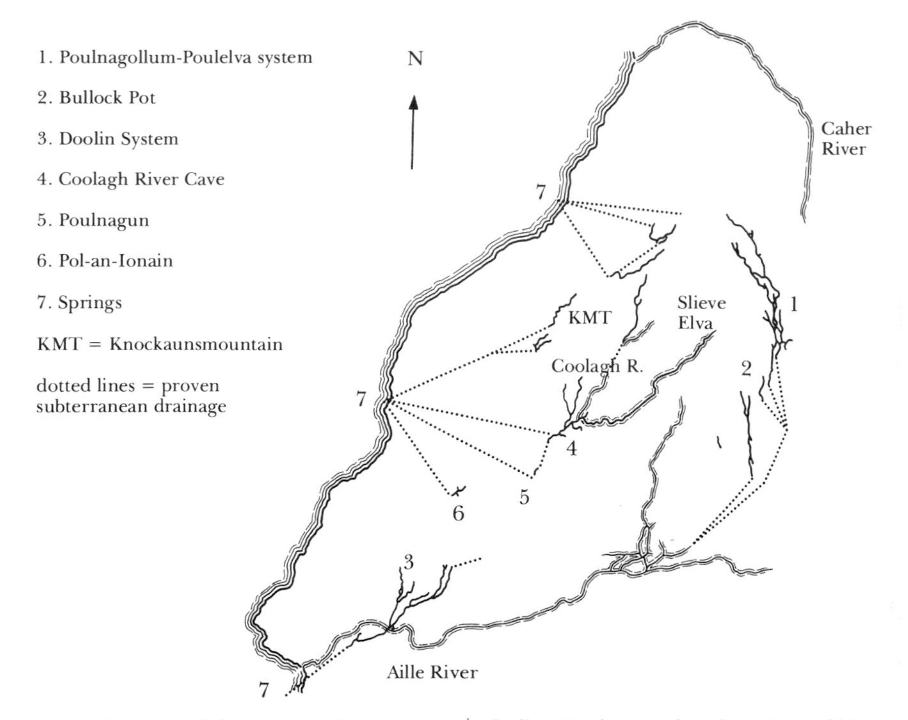

Fig. 3.20 Caves of the Western Burren, not including Poulnacapple (after C.A. Self *Caves of County Clare* University of Bristol Spelaeological Society).

Most of the major caves are found on the flanks of the hills north of Lisdoonvarna. Small streams seep from the summit peat bogs to drain radially down the hillsides until they meet the limestone. Here they go underground immediately and, with few exceptions, flow down-dip towards the south in simple vadose stream passages. For many cavers, the sinuous underground canyons with their glistening water-sculpted walls echoing the sound of rushing water, represent the nearest thing to paradise on earth.

The longest cave in Ireland starts up on the eastern flank of Slieve Elva, the highest of the hills. Poulnagollum – 'The Cave of the Doves' – contains almost 15 km of passages and was the first of the Co. Clare stream caves to be explored. It was an obvious site for investigation, as the deep, cliff-walled entrance hole contains a substantial stream which appears from a cave (Upper Poulnagollum, fed by the shale-edge swallets) and promptly disappears off into another cave (the Main Streamway). The earliest reference to exploration of the cave is a chalk inscription, dated 7.10.1880, made by W.H.S. Westropp and party, although anecdotal evidence from a geological journal of the time suggests that members of the Irish Geological Survey may have already visited the cave "having fortified their nerves with liberal allowances of whiskey." Mapping of the cave was begun by E.A. Baker, the well-known English caver, in 1925, but the main work was done between 1941 and 1943 by J.C. Coleman and N.J. Dunnington. Their classic paper appeared in the *Proceedings of the Royal Irish Academy* and excited much interest in England, particularly among members of the University of Bristol Spelaeological Society. The UBSS had soon begun a collaboration with Coleman in Poulnagollum and this quickly blossomed into a love affair with the landscape and caves of Co. Clare. Inspired by the enthusiasm and dedication of Prof.E.K. Tratman and Dr.O.C. Lloyd, both of whom continued to go underground in Clare well into their seventies, UBSS members over a 40 year plus period systematically pieced together a detailed picture of the area's caves, culminating in the production of two notable books: E.K. Tratman's *The Caves of North-West Clare, Ireland* (David & Charles, 1969), and C.A. Self's *Caves of County Clare* (UBSS, 1981).

Poulnagollum is a beautiful cave, unusually complex for Co. Clare. A subterranean link with the Poulelva pothole provides one of the easiest, classic caving through-trips in the British Isles. Hydrological connections with the difficult caves E2 and Bullock Pot, at opposite ends of the system, hold out the prospect of a much longer, but extremely nasty through-trip, if caver-sized links are ever found. The streamways of Upper Poulnagollum and Upper Poulelva are in many ways alike, and are typical of the area. Red, peaty water burbles along, ankle-deep, in clean-washed, T–shaped canyon passages with scalloped walls. The arms of the 'T' formed first, under phreatic conditions. As the passage enlarged, it was unable to dissolve its floor owing to the presence of an insoluble band of chert, and instead developed laterally, taking a sinuous, braided course which shows in the arms of the 'T', as *anastomosed* roof pendants and loops. Later, a lowered water-table caused the stream to develop an air-space and it began to abrade its chert bed with teeth of grit, eventually slicing through to form the stem of the 'T'. Where the thicker chert beds were breached, underground waterfalls resulted, and at the Poulnagollum and Poulelva potholes, the inlet streams cascade down a series of chert-topped chutes.

In the upper streamways, later speleothem deposits have begun to embellish and confuse the clean, water-carved contours. The stubby stalactites which pepper the cave roof are garlanded with strands of grass brought in by floods. The canyon passages are still very young – formed since the last glacial period, and are in a stage of rapid development.

The Main Streamway below Poulnagollum Pot is a handsome passage; 10 m tall by a metre wide, it sweeps southwards in a series of sinuous curves. Because subterranean meanders, like those of surface rivers, migrate downstream, the upper and lower levels of the canyon are often completely out of phase, with deep undercuts, protruding tongues of rock and scalloped slip-off slopes. This wonderful streamway gathers tributaries from the east, such as Branch Passage – an active inlet which can be followed for over three kilometres upstream, and the dry, fossil High Road which formed before the last glaciation.

At the southern end of Mainstream Passage, the cave becomes wider and lower and enters a confusing region called The Maze, where the linking passage from Poulelva enters from the west. The stream has by now been lost, but is seen briefly again to the west as it sneaks through immature, flooded passages on its way to Bullock Pot. A little south of Bullock Pot, the same water appears briefly on the surface before vanishing into an immature cave which conducts it as far as Lisdoonvarna. Here it resurges in the guise of the Gowlaun River, a tributary of the Aille River which flows across the shale bridge then over till-covered limestone to the sea.

Clare's number two cave, the ten kilometre-long Doolin System, is in many ways similar to Poulnagollum. It is a dendritic system with large active streamways and older, dry, 'fossil' passages. Its catchment is the south-western part of the shale bridge and it drains to an intertidal resurgence buried beneath a sandy beach. Doolin Cave is in the early stages of capturing the Poulnagollum cave water, in a rather curious manner. The Aille River, carrying the Poulnagollum waters on their way to the sea, crosses over the top of Doolin Cave. Barely six metres thickness of limestone and a thin layer of glacial till separate the surface river bed from the cave ceiling, and water leaks from the one to the other at the spectacular Aille Cascade.

Without doubt, the most exciting cave of the region is the five kilometre-long Coolagh River Cave, with its huge catchment comprising most of western Slieve Elva, eastern Knockaunsmountain and the northern part of the shale bridge. The principal sink, Polldonough, is in a karst window within the shale bridge, opened by the downcutting of the Coolagh River. Beyond the sink, the dry valley continues south-west over a saddle, and is joined by a large tributary stream which sinks at the Polldonough South swallet.

Coolagh River Cave has all the attributes which typify a youthful Clare cave, but with added water – there is no escaping it, whether in the low, wet bedding passages, or in the large Main Drain, formed where the Polldonough and Polldonough South streamways join their considerable forces. Here, the remains of a small tree or two casually wedged in the roof hint that this could be a cave which floods with a capital 'F'. At the lower end of the Main Drain, two more tributaries join the cave, one from the long Polldonough North branch, the

Fig. 3.21 The flood-prone main passage of the Coolagh River Cave can fill to the roof following heavy rain. (Paul Deakin)

Fig. 3.22 The Great Stalactite in Pol-an-Ionain, claimed to be the world's largest (Paul Deakin)

other from Pollclabber. The even larger streamway now enters a sump, where its progress is impeded by underwater mud banks.

Up on the surface, the dry Coolagh Valley continues south and west as a limestone-floored ribbon flanked by hills of shale. In the valley bottom, two potholes provide glimpses of the underground river: Poll Cloghaun is an enlarged joint; Poulnagun is a deep hole in the shales of the valley flank, with limestone exposed at its base. Further on, the valley reaches its most spectacular as it passes beneath Ballynalackan Castle, picturesquely posed on a white limestone cliff, until suddenly, at the farmhouse of Cregg Lodge, it begins to dwindle away, becoming shallower and finally petering out completely on the coastal limestone platform.

Back near the castle, a tiny stream sinks into the crawly cave passage of Pol-an-Ionain. After a miserable 150 m, the visitor pops out into a large chamber, an isolated remnant of a pre-glacial cave now choked with sediments. In the centre of the ceiling hangs a massive stalactite, 6.5 m long, claimed by the *Guinness Book of Records* to be the largest of its kind in the world.

The thrill of a trip into the Main Drain of the Coolagh River Cave places it high on the itinerary for cavers visiting the area. Only two people, however, have seen the cave in 'wet' conditions. In August 1950, Kay Dixon and Johnny Pitts entered the cave via Polldonough South to retrieve equipment left behind after a long surveying trip the previous day. One of the peculiar features of the Coolagh River Cave, not known in 1950, is that it becomes impassable near the entrances and then fills up from the bottom (due to the mudbank constrictions in the terminal sump). When Pitts returned to the bedding cave canal, almost in sight of the entrance, the force of the water washed him back into the cave. He and Dixon were forced to sit out the flood in the roof of nearby Double Passage. The low point of their morale came when the cold air being drawn in with the water suddenly stopped – the cave entrance was underwater. Fortunately this was only a small flood and the water fell again as fast as it had risen.

A much larger deluge occurred during a storm in August 1967, submerging Polldonough and Polldonough South to a depth of seven metres. The cave could not cope with the flow and a river fountained out of the ground from

Fig. 3.23 A cave diver in one of the Green Holes of the Hell Complex off Doolin Point. The rock walls are encrusted with marine life, including sponges, anemones and sea urchins. (Michael Pitts)

cracks in the valley floor above the sump, carrying flakes of shale blasted from the stream bed 40 m below. The flood then careered down the valley past Ballynalackan Castle to Cregg Lodge Farmhouse, where it was swallowed by a tiny cave which has been dye-traced to a series of resurgence cracks on the sea shore north-west from Cregg Lodge (an entirely unexpected direction). A submarine resurgence in the same bay discharges peaty water when the caves are in flood.

In Clare, caves which have been drowned by the sea are called 'Green Holes'. The most famous occur in the sloping limestone platforms of Doolin Point. At a depth of around 13 m, the rocks are riddled with caves which inter-connect with each other and with a land-locked open joint in whose depths the sea seethes and boils during stormy weather, giving rise to its local name 'Hell'. The premier discovery to date is situated in a small cove 200 m north-east of the Point. Mermaid's Hole, at 10 m depth, can be followed from salt- into fresh water. Cave divers have now pushed inland along two flooded passages requiring dives respectively of 450 and 700 m. This is an outstanding achievement, since diving is only possible during slack-tide and then only when the sea is calm.

North of the Green Holes, a number of small caves open into the intertidal zone. Known collectively as the Brown Holes, they appear to be similar, if smaller, versions of the Green Holes. From the gentle southerly dip of the strata, they may even be developed in the same bed of rock.

Mermaid's Hole is the only one of these caves which is currently active. It carries a tiny freshwater stream fed from the coastal platform. Although the cave has not yet been studied in detail, some features noted by the divers suggest that Mermaid's and other Green and Brown Holes could be ancient distributary resurgences, possibly related to a pre-glacial cave in the Doolin or the Coolagh River valley.

Other Areas

Highland

Perhaps the best known, and certainly the most photographed cave in Scotland is the huge Smoo Cave which opens at the base of towering cliffs to the east of Durness. Originally formed by solution of the ancient Cambrian Durness Limestone, it has been enlarged by wave action to form a chamber measuring about 40 m across by 20 m high, stretching 60 m or so back into the cliff. The recent nearby discovery of submarine caves, similar to the Doolin Green Holes, is of particular interest to cave-diving naturalists.

Near Ichnadamph in the Assynt region of north-west Sutherland lies the pretty lochan of an Claonaite. The lake is fed by a number of burns, but none flows out. Its waters disappear into the surrounding limestone, reappearing at the foot of the glen as 'Fuarain Allt nan Uamh' – the Great Spring of the Burn of the Caves. In between lie ten, mostly short, caves, including the main drain, Uamh an Claonaite – a 'cavers' cave' – wet, cold and arduous. Above the streamway lies a surprising series of vast fossil tunnels, almost blocked by rolling dunes of sand and stretching for over a kilometre.

The island of Skye also contains a small amount of cavernous Durness Limestone. Of the 50 or so caves recorded, the 350 metre-long Uamh an Ard Achadh (High Pasture Cave) is the most extensive. Several of the caves carry

Fig. 3.24 The impressive entrance of Smoo Cave, developed in the Cambrian Durness Limestone. (Paul Deakin)

streams which originate on Beinn an Dubhaich, a granite intrusion which split the limestone outcrop in the Tertiary, causing cracks which became filled with igneous dykes. These have influenced the development of caves such as Camas Malag, whose stream is briefly forced to the surface by a dyke before sinking back into the limestone.

Gwent and Gloucestershire

Otter Hole is a stream cave which resurges on the west bank of the River Wye near Chepstow. An upper abandoned series of passages, spectacularly decorated with speleothems, and the lower streamway, appear to be related to the lowest terraces of the Wye Valley, while a tidal sump in the entrance series is the result of a post-glacial rise in sea level. With a passage length of 3200 m, Otter Hole is the longest cave in the Carboniferous Limestones of Gwent, though the Forest of Dean to the east of the Wye contains many smaller caves.

Fig. 3.25 The Hall of the Thirty, Otter Hole – one of the most lavishly decorated caves in Britain. The cave runs beneath the Chepstow Racecourse to the bank of the River Wye. (Clive Westlake)

Devon

The most extensive of the Devon cave systems is situated close to the southeast margin of Dartmoor, in the Devonian limestones exposed along the valley of the River Dart near the village of Buckfastleigh. Joint-Mitnor Cave contains the richest deposit of interglacial mammalian remains ever found in Britain.

Nearby Rift Cave, Reed's Cave and Baker's Pit originally formed part of the same hydrological system, but the latter alone remains active, the others having been drained long ago. Joint-Mitnor, Rift and Reed's Caves are important roosting and nursery sites for Greater and Lesser Horseshoe Bats. To the north of Buckfastleigh Hill lies the intricate network of Pridhamsleigh Cavern.

A biologically and geologically-interesting group of caves occurs on Berry Head, close to Brixham. They have an unusual morphology, suggesting that they were formed at the base of a thin freshwater lens overlying a deeper seawater *aquifer*, in a manner analogous to the Blue Holes of the Bahamas. Starfish, Cuttlefish and Garfish Caves, as their names suggest, open in the intertidal zone of the sea-shore and contain some interesting marine life, discussed later in Chapter 5.

Caves in Cretaceous Chalk

The 366 metre-long Beachy Head Cave near Eastbourne in Sussex opens part-way up the cliff about half a kilometre west of the lighthouse. It is by far the longest and roomiest cave so far explored in Cretaceous chalk in the British Isles, with a passage diameter approaching two metres in places, although most of the cave is of crawling size. Other chalk caves occur around Beer in East Devon. Such caves are of interest because they demonstrate that drainage in the chalk follows a very similar pattern to that in limestone, although on a smaller scale because of the much lower mechanical strength of the former.

Landslip Caves and Mines in Jurassic Limestone

In Britain, the most extensive landslip caves, also known as 'gulls', occur in a belt of Jurassic rocks stretching diagonally from Teeside to the Dorset coast. They are formed by downhill sliding within inclined strata, producing rift-like cave passages. Some gulls are very old; Sally's Rift near Bathford, Avon has been dated beyond 350,000 years (by the Uranium/Thorium ratio within the rock, which acts as an atomic clock). The Hambleton Hills of North Yorkshire, the Cotswolds and the Isle of Portland in Dorset all have significant slip caves, some extending hundreds of metres.

The Jurassic strata have produced some of England's finest building stone, much of which was extracted by mining. Some abandoned Bathstone and Portland Stone mines are particularly extensive and provide a rich hunting ground for the cave naturalist and bat recorder.

Co. Mayo and Co. Galway

Cong is a low lying isthmus of Carboniferous limestone no more than three and a half kilometres wide sandwiched between Lough Mask and Lough Corrib on the Galway / Mayo border. It is an area of complex, closed depressions, dry valleys and limestone pavement overlain by glacial drift. The waters of Lough Mask cross the Cong isthmus to Lough Corrib via a series of wet underground, joint-controlled and bedding-plane passages. Surface fissures give access to short sections of the complex, though the longest cave of the area, Ballymaglancy, appears to belong to an unrelated hydrological system.

The Aille River receives the outflow from Glenmask and water shed by the Partry mountains south of Westport. After a course of 15 km it leaves the sandstones and meets the thickly bedded limestone at Aille. Here it sinks underground into the Aille River Cave, a complex network of floodprone passages which offers sporting caving and a series of intimidating deep dives.

The limestone around the town of Gort is riddled with impenetrable flooded caves. The River Beagh flows out of Lough Cutra and sinks after three kilometres into fissured limestone, reappearing in a wide depression called the Punchbowl. Thereafter it can be seen in a number of impenetrable 'eyeholes'. At Coole it rises and sinks repeatedly in shallow beds on its way into Coole Lough.

Co. Antrim

Perhaps the most unlikely Irish discovery is the 450 m of unique passage in the brilliant white chalky limestone of the Antrim plateau above Carnlough. It can only be explored at times of exceptional drought as the entrance lies in the bed of the Blackburn stream. So far it has been pushed to Sump 5 in pleasing phreatic tubes, the walls punctuated with nodules of flint.

Co. Kerry

The Castleisland / Tralee area has in recent years become a prominent caving region. Crag Cave contains over four kilometres of beautifully decorated passage, part of which has been opened as a fine show cave. It is an old system with a series of well-decorated fossil levels perched above the modern streamway.

Clwyd

North Wales boasts two significant caving areas – along the gorge of the River Alyn, and around the lead mining village of Minera, a few kilometers west of Wrexham.

Ogof Hesp Alyn lay entirely below the water table until the early part of this century when it was drained by mining activity, offering non-diving cavers a rare chance to explore an unmodified phreatic cave system. Over two kilometres of extremely muddy passages have so far been explored, although few cavers have visited the further reaches of the cave more than once!

The Minera caves constitute the major system in North Wales. The active caves Ogof Llyn Parc and Ogof LLyn Ddu, together with the finely-decorated fossil system of Ogof Dydd Byraf, together contain over eight kilometres of passages connecting with old mine workings which in some cases provide the only access.

4

Cave Fauna and Flora

The origin and ecological classification of cavernicoles

Since the early part of this century, it has been traditional for biologists to classify animals found in caves on the basis of their degree of ecological dependence on the cave habitat, in what is called the Schiner-Racovitza system, after its originator and subsequent modifier (Schiner, 1854; Racovitza, 1907). The system recognizes three main categories:

'Trogloxenes', or cave visitors, occur in caves but do not complete their whole life cycle there. They are usually subdivided into 'accidental trogloxenes' which have been carried passively into the cave by gravity, flowing water, or other agencies, and 'habitual trogloxenes' (such as bats) which seek shelter in the cave, but leave it to feed, or which visit the cave at certain times in their life cycle.

'Troglophiles', or cave lovers, are facultative cavernicoles. They are found living permanently, and successfully complete their life cycles, in caves; but they also do this in suitable non-cave habitats.

'Troglobites', or cave dwellers, are obligate cavernicoles which cannot survive outside caves.

Albert Vandel, in his monograph *Biospeology: The Biology of Cavernicolous Animals*, published in 1964, states that:

> "The true cavernicoles, the troglobites, are for the most part relicts ... They (at least the terrestrial forms) are found only in Europe, North America, and Japan, regions which underwent marked climatic modifications during the later geological periods ... The subterranean world represents a refuge which has allowed animal forms which have been driven from the surface of the earth by climatic alterations, to persist to the present day."

This theme has been taken up by most cave biologists and, until recently, the concept of troglobites as 'relicts' or 'living fossils' was so deeply ingrained, that 'relictualness' was used by many systematists as an unofficial parameter in defining newly-discovered cavernicoles either as 'troglobites' or as 'troglophiles'. However, in the last twenty years, it has emerged that Vandel's "true cavernicoles" are in fact not by any means confined to Europe, north America and Japan, but have a world-wide occurrence. Furthermore, many obligate cave species are not relicts, but are recently recruited from actively speciating taxa which inhabit regions in which the caves themselves are still in the process of developing.

New species arise for the most part as the result of novel selection pressures acting on the genotype of an isolated population of a species which is in process of expanding its range. This phenomenon is particularly well illustrated on oceanic islands (the classic example being the Galapagos), but may be equally marked on 'ecological islands' within a single landmass. Caves (systems of

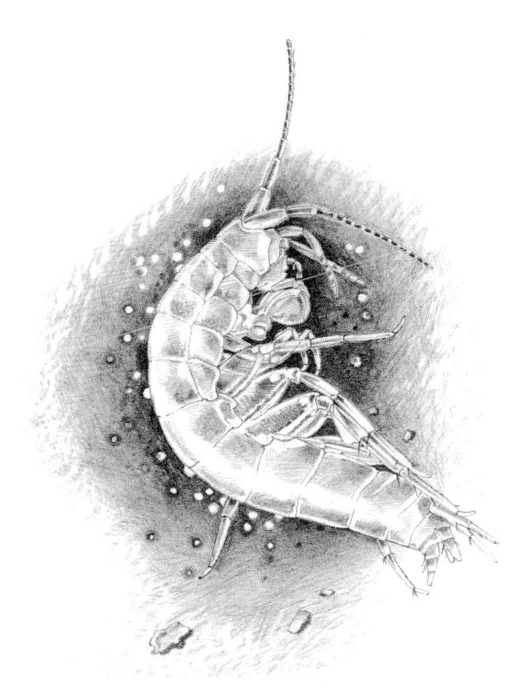

Fig. 4.1 The amphipod *Niphargus glenneii* is a classic troglobite – eyeless, entirely without pigment and confined to subterranean waters in the limestone caves of Devon.

interconnected, dark, rock-bound voids) in some ways resemble islands, or archipelagos, in that they constitute a discrete and alien environment with more or less defined limits. In some cases they are marooned in a 'sea' of non-cavernous rocks, in others linked to other cavernous complexes by rocks with a sufficiently developed system of sub-soil fractures (SUC) to allow limited dispersal from one cave 'island' to the next. Terrestrial limestone cave habitats tend to be well integrated immediately beneath the soil, but less well integrated at greater depths, so that the SUC-specialized cavernicoles tend to have wider geographical distributions than 'deep cave' specialists. The same is true of aquatic cavernicoles, where species inhabiting the most superficial groundwaters (such as the amphipods *Bactrurus mucronatus* in North America, and our own *Niphargus aquilex*) may achieve far more rapid rates of spread than relatives which are confined to the deep phreas of caves.

There would seem, *a priori*, to be a greater chance of speciation events occurring in the depths of deep limestone cave systems (the most remote islands of the karst archipelagos) than in almost any other habitat in the British Isles. Why then are our caves not filled with scores of unique, endemic troglobites?

Perhaps they once were, but if so, the ice ages of the Pleistocene have left few survivors (but see Chapter 6). The damage was done over a period of 1.7 million years by a sequence of around 17 alternating glacial and interglacial periods which changed the landscape of our islands forever. Those small, isolated karst areas which escaped glaciation and flooding – parts of Devon, Dorset, Mendip and the South Downs – provided relatively little habitable space for cavernicoles when compared with the huge expanse of well-developed karst in southern Europe which survived the period unscathed. It is the fauna of the latter area which has since spread outwards to recolonize and diversify in the adjacent continental landmass. Our own smaller and relatively less-cavernous refugia seem to have retained little more than a few groundwater Crustacea from pre-Pleistocene times, so that many of the niches which in Europe are occupied by highly modified ancient troglobites, are taken here by relatively recent colonists. The majority of these may well inhabit mesocavernous rock-crevices in preference to larger cave passages. As such habitats become better investigated and with increased understanding of the ancestry and kinship affinities within groups such as the springtails, mites and ostracods, I predict that quite a few more ecologically cave-specialized species may be recognized within the British fauna.

The problem with a system of classification based on ecological or distributional criteria is that it requires the collection of a vast amount of data, before it can be used with any confidence. Fortunately for naturalists in Britain and Ireland, good distributional data are available for many invertebrate taxa so that it should (at least in theory) be possible to discover from an exhaustive search of the literature which of our cavernicoles have been recorded from non-cave habitats. It is then necessary to decide whether their presence outside of caves is facultative, or accidental (aquatic troglobites may be accidentally washed out into springs, or flushed into lakes, where they may survive for a time).

In countries where the distributions of invertebrate faunas are not so well documented (this applies to most of the world), it may be impossible to distinguish between 'troglobites' and 'troglophiles' with any degree of accuracy – at least for the foreseeable future. In recognition of this, most cave biologists have adopted an unofficial set of *morphological* criteria which are used to distinguish between 'troglobites' and 'troglophiles'. The criteria are based on the fact that many well-studied cavernicoles whose distributions seem to be strictly 'troglobitic', share in common a high degree of depigmentation and eye-regression. It is thought that such features may become established only in populations whose members are confined underground, but will be weeded out from the gene pool of troglophile populations, since they are clearly counter-adaptive in habitats which are not perpetually dark. In consequence, it has become the convention among most cave biologists to reserve the term 'troglobite' exclusively for species with regressed eyes and strongly reduced pigmentation. Unfortunately, eyelessness and depigmentation are features equally common among specialized soil-dwellers or interstitial organisms as among cavernicoles, so that, without adequate distributional data it is still almost impossible to recognize tropical 'troglobites' as such. Culver (1982), writing about temperate American caves, states that many 'troglophiles' are so classified

"only because they show little sign of regressive evolution. Other troglophiles have no surface populations near cave populations. For example *Gyrinophilus porphyriticus* [a salamander] is common in caves in the upper Powell Valley in Virginia and Tennesse, but no suface populations are known from this area ... It is very rare to find a surface population of any cave organism that extends from the surface directly into the cave. In most

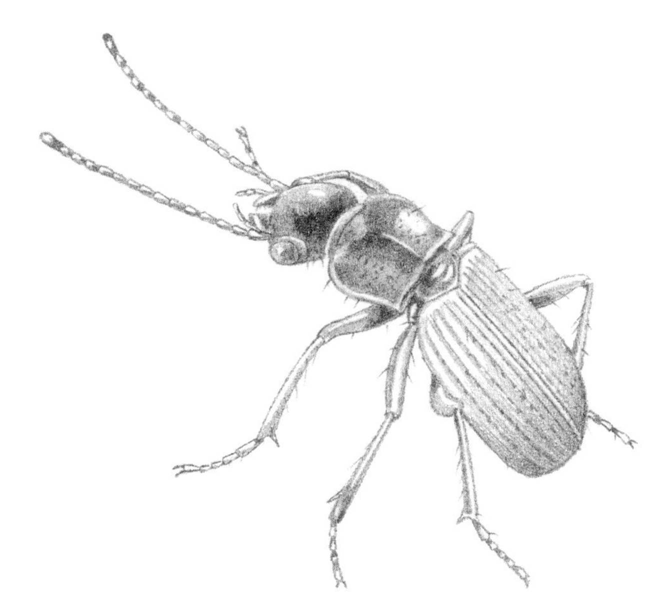

Fig. 4.2 The ground beetle, *Trechus micros*, is neither eyeless, nor endowed with particularly long legs and antennae. Cave biologists have therefore tended to classify it as a troglophile, or facultative cave inhabitant. Yet it seems to live only in cave or SUC habitats in Britain and is really a successful caverniate.

cases it is unlikely that gene flow is retarding adaptation. It is likely that what we call 'troglophiles' have been in caves for a shorter period of time than what we call 'troglobites', if only because troglophiles show less regressive evolution ... Nonetheless, by any measure available, many troglophiles are very successful in caves."

It is by no means clear whether *anophthalmia* or depigmentation have any great adaptive significance for troglobites, nor how quickly such features become fixed in cavernicolous populations, and it would therefore seem unwise to rely too heavily on their presence as the sole admissible evidence of cave-dependency. In our own caves, many of which remained ice-covered and probably uninhabitable, until as recently as the last (Devonsian) glaciation, a mere 12,000 years ago, it is likely that a number of troglobites (as defined strictly in the Schiner-Racovitza system) will have become isolated in caves too recently for them to show typical 'troglobitic' modifications to the extent characteristic of more ancient troglobites found in the caves of southern Europe.

In the same way that the Hawaiian mesocavernicoles (discussed in the last chapter) make periodic sorties into lava caves to exploit the resources within, so other cavernicoles may venture out into the cave entrance, or beyond, on moist, still nights, in order to feed. In Britain, non-cave habitats (such as soil), may receive far more study than caves, so that straying cavernicoles may be recorded with disproportionate frequency, giving the illusion that British cavernicoles are less cave-dependent than they really are. In tropical karst areas, the fauna of caves may have received more study than that of related non-cave habitats, producing the opposite illusion – that some invertebrate species are more cave-dependent than is really the case. What this adds up to, is that there are considerable problems in correctly assigning any cavernicole (other than an ancient, morphologically-specialized troglobite) to its correct category within the Schiner-Racovitza system.

I am sure that some colleagues would claim that in drawing attention to the existence of a distinctive suite of specialized subterranean animals, the Schiner-Racovitza system has helped to stimulate interest in caves among the biological community as a whole and so has been a good thing for the discipline of cave biology. I would argue that, in its current usage, it has had two unfortunate effects: firstly, it has focussed attention on the non-adaptive, or regressive features of certain ancient cavernicoles rather than on those adaptive features which allow *all* successful cavernicoles to cope with the combined problems of perpetual darkness, low, or sporadic food availability, and a rigorous climate. Secondly, the identification of morphologically-defined 'troglobites' as the 'true' cave inhabitants has led to a dismissal of the British and Irish cave faunas as being of relatively little interest. This is an unfortunate and quite misguided conclusion. Our cave fauna is in a dynamic phase of colonization and adaptation brought about by the extinction of much of its original complement of ancient 'troglobites' during the climatic upheavals of the Pleistocene. There is thus enormous scope in our caves for research into the process of cavernicolous evolution.

In 1895, an Australian zoologist, A. Dendy, proposed the term 'cryptozoa' (meaning hidden animals) to characterize creatures found dwelling in darkness beneath stones, rotten logs, tree bark and similar places. Recently, leading tropical cave biologist Stewart Peck has suggested that the term be extended to include all the main categories of organisms "which live hidden lives in spaces at or beneath the earth's surface", including the inhabitants of soil, ground litter and caves. With the proviso that the term be widened to the more correct 'cryptobiont' (plural 'cryptobia'), I feel that Peck's is a positive suggestion which will increasingly find favour with other cave biologists and will result in more cross-fertilization of ideas between forest, soil and cave ecologists and a greater interest in caves among the mainstream of biological researchers.

Because this book is specifically about caves, I do not propose to use the wider term 'cryptobiont' in the following pages. But neither do I propose to labour the narrow distinction between 'troglophile' and 'troglobite'. Instead, I shall in general use the perfectly adequate term 'cavernicole' for a species which lives in caves and can complete its life cycle there, though it may live in other 'cryptobiotic' habitats as well. I will, however, retain the term 'troglobite' in its restricted latter-day sense as 'a cavernicole which shows morphological features which point to a long history of cryptobiotic adaptation'. With this definition, it is not necessary to establish empirically that a 'troglobite' is exclusively confined to the cave habitat.

Life on the threshold

Green plants

The cave threshold is often sheltered from wind and direct sunlight, with less variation in temperature and humidity than is found in the world outside. The gradual decrease of light intensity with increased distance from the entrance brings about a distinct zonation of the plants, with increasingly shade-tolerant, moisture-dependent species succeeding each other to a point where light intensity becomes too low to support photosynthesis. This physiological succession is mirrored in the taxonomic make-up of the various plant zones. Flowering plants near the entrance are replaced first by ferns, then mosses, then liverworts, until the dimness of the light will support only green algae, and finally, only cyanophytes (better known as "blue-green algae").

Typical flowering plants of cave entrances include Ivy (*Hedera helix*), Herb Robert (*Geranium robertianum*), Wood Sorrel (*Oxalis acetosella*), Dog's Mercury (*Mercurialis perennis*), Cuckoo Pint *(Arum maculatum)*, Stinging Nettle *(Urtica dioica)* and Dog-violet (*Viola riviniana*). The latter has a distinct deep-shade form, with very much enlarged leaves. A similar phenomenon occurs in deep-threshold individuals of the mosses *Eucladium verticillatum, Adoxa* and *Phyllitis* spp., and the fern genus *Asplenium*, where enlarged leaves are presumably a simple growth response to counter the poor photosynthetic performance of the plant at low light levels. The liverwort *Conocephalum conicum*, a very common species in cave entrances, produces a stunted form at the innermost limit of growth, and young fern prothalli (the gamete-producing stages) are unable to develop sexual organs at very low light levels.

The above list also contains at least one plant species (*Oxalis acetosella*) which is generally associated with acid, rather than basic soils. This is because cave entrances often accumulate deep layers of soil and organic material, which may produce locally acid conditions.

In the Western Burren of Co. Clare, almost every entrance to the Cullaun and Doolin caves is filled with a 3–10 m high thicket of the Goat Willow or Great Sallow, *Salix caprea*, whose bright yellow 'pussy willow' catkins provide a feast for the bees in early spring. Like most members of the willow family,

Fig. 4.3 Thickets of sallow, *Salix caprea*, are a characteristic feature of cave entrances on the Burren of Co. Clare. Their usefulness as landmarks for cave seekers has earned them the nickname "Spelaeodendron" in some caving circles.

S. caprea prefers damp or marshy land and a good water supply. Such conditions are relatively rare on the bare, ice-scoured limestone pavements of the Burren. However, during the ice ages, the hollows containing the entrances to the underlying river caves became filled with shale debris or glacial till which has since weathered to produce a rich soil, providing perfect conditions for the trees to flourish. The association between *S. caprea* and caves proved so helpful to the early cave explorers that the tree earned the nickname 'Spelaeodendron'.

A number of mosses are found in cave entrances. One of the commonest is *Gymnostomum aeruginosum*, which forms dense, compact, deep green tufts or cushions in damp areas of cave mouths and which may become more or less encrusted with calcite deposits. Another cushion-forming species, *Eucladium verticillatum*, together with *Cratoneuron filicinum*, a golden-green, pinnately-branched moss, prefer even wetter habitats and often grow in flowing springs where they may provide the scaffolding for tufa dams. The tall, leafy, light-green fronds of *Mnium undulatum* are to be found lurking in dingy corners of Yorkshire potholes, and the intricate, brownish-green mats of *Heterocladium heteropterum* can be seen in the seaside caves of the Gower Peninsula.

The most crepuscular, and one of the most beautiful of all our mosses is *Schistostega pennata*, which grows in the dimmest recesses of sandstone cracks and caves, old Cornish mine shafts, and even down rabbit burrows, but avoids the base-rich soils of equivalent limestone habitats. *Schistostega* forms wide, loose carpets of slender shoots which grow 5–12 cm high and are reminiscent of tiny Polypody ferns. The protonema of the leaves has a peculiar light-reflecting power which produces a kind of starry effect, so that the plants seem to flood the dark crevices where they grow with a luminous, golden-green light.

Among a wide range of ferns recorded from cave entrances, several are common enough to be considered 'typical' of this habitat. They are: Hart's Tongue Fern (*Phyllitis scolopendrium*) with overwintering tufts of strap-shaped leaves; Maidenhair Spleenwort (*Asplenium trichomanes*) with narrow, wiry fronds divided into tiny oblong leaflets; Wall-rue (*A. ruta-muraria*) which forms small, drooping clumps of irregular fronds; and Rustyback Fern (*Ceterach officinarum*) with dense tufts of dark, pinnately-lobed fronds backed by rust-brown scales. Four upright 'feathery' ferns can also be added to this list: the Male Fern (*Dryopteris filix-mas*), Hard Shield Fern (*Polystichum aculeatum*), Lady Fern (*Athyrium filix-femina*) and Brittle Bladder Fern (*Cystopteris fragilis*). The latter species is common in the north of England, particularly in cave entrances in the Yorkshire Dales, but until recently was considered rare in the south. About fifteen years ago, the first colony of Brittle Bladder Ferns was reported in a Mendip cave entrance, and today several caves in Somerset have established colonies. Charlie Self, who pointed out this phenomenon to me, suggests a link with the opening of the M5 motorway some 20 years ago, which for the first time allowed Mendip and Yorkshire cavers easy access to each others' areas at weekends. Cavers, being a somewhat scruffy breed, may have accidentally introduced the spores to Mendip on their clothes and boots. With the increased participation of British cavers in foreign expeditions, it will be interesting to see if even more exotic introductions occur in the future.

Other species which occur regularly are: the ubiquitous Bracken (*Pteridium aquilinum*); Common Polypody (*Polypodium vulgare*) with single, more elegant singly-pinnate fronds; Limestone Fern (*Gymnocarpium robertianum*) with

Fig. 4.4 Brittle Bladder Fern, *Cystopteris fragilis*. Originally confined to the north, it has recently spread to Mendip. Visiting cavers may have helped by bringing in spores on their muddy boots.

bracken-like (but smaller) three-pinnate fronds; Green Spleenwort (*Asplenium viride*), a close relative and look-alike of Maidenhair Spleenwort; and the feathery Rigid Buckler Fern *(Dryopteris submontana)*.

Western Black Spleenwort (*A. onopteris*), a delicate feathery fern, appears to be associated with limestone only in Ireland, while a couple of ferns are notable for their association with sea caves in Britain. The common Sea Spleenwort (*Asplenium marinum*) is found on cliffs and in damp sea caves mainly on western coasts, and also in the grikes of limestone pavements on the Burren, while the very rare Dickie's Fern (*Cystopteris dickieana*) is known in Britain only from one or two sea caves south of Aberdeen.

The Maidenhair Fern (*Adiantum capillus-veneris*) has been found growing close to electric lights installed by the owners of Wookey Hole show cave in 'Chamber 8', well into the cave system. This delicate fern, with its fan-shaped pinnules held on slender dark stalks, is seldom found inland in Britain as it is intolerant of frost. It is therefore interesting to speculate how the fern spores can have reached such an isolated spot – cut off upstream by the phreatic River Axe, and linked to the entrance part of the cave by an artificial tunnel. Hart's Tongue and Rustyback ferns are also commonly found growing around show cave lights.

Plants which grow around artificial lights in caves (known collectively by the German term '*lampenflora*') often show a similar zonation to that found at cave entrances. They are usually unwelcome to the cave owner, as they are felt to detract from the alien atmosphere which visitors expect from a show cave, and because they discolour calcite formations. The shade-tolerant blue-green alga *Anacystis montana* is a common culprit in British caves.

A decade or so ago, the painted cave of Lascaux in the Dordogne, France, was the site of a heavy growth of a green alga, *Palmellococcus*, which threatened to overgrow the prehistoric paintings. This 'Maladie Verte' was the subject of much research and was eventually controlled by application of formalin solutions to kill the algae, following treatment with a mixture of antibiotics to kill the bacteria which were thought to be enriching the nutrients available to *Palmellococcus* on the walls. The source of these nutrients was ascribed to "organic contamination derived from the large number of visitors". According to Cubbon (1976), green algae growing in the dark can retain their chlorophyll, hence the colour of the 'Maladie', although of course the cave was illuminated for visiting parties.

Some algae and cyanophytes appear to be highly specialized for life at the low light levels of the deep threshold. Australian botanists Cox and Marchant reported in 1977 that a Chlorococcalean alga, *Chlorella* sp., and a Chroococcalean cyanophyte, *Chroococcus* sp., from a cave in Australia, showed an exceptionally high thylakoid density (thylakoids being the structures which intercept light). This finding has since been confirmed in a number of other deep-threshold species, although none from British or Irish caves.

Another phenomenon sometimes met with in thresholds is the formation of 'eucladioliths', which are tufa-like deposits formed when seepage water runs over growing moss plants (generally *Eucladium* and *Gymnostomum*). As the moss grows towards the light, more lime is deposited on the older parts, and a formation builds up as a tube around the plant. A similar process occurs with growing liverworts, algal filaments and cyanobacteria. The explanation would seem to be that removal of CO_2 (carbon dioxide) by the plants during photosynthesis causes a shift in the chemical balance which maintains $CaCO_3$ (calcium carbonate) in solution, causing the excess to precipitate out. Stalactites formed around colonies of the alga *Gleocystis rupestris* have been found in the deep threshold region of a cave in Germany by Dobat (1971). These too

Fig. 4.5 Toothwort, *Lathraea squamaria*, a parasite on tree roots, is one of the few higher plants able to grow successfully in the sunless depths of caves.

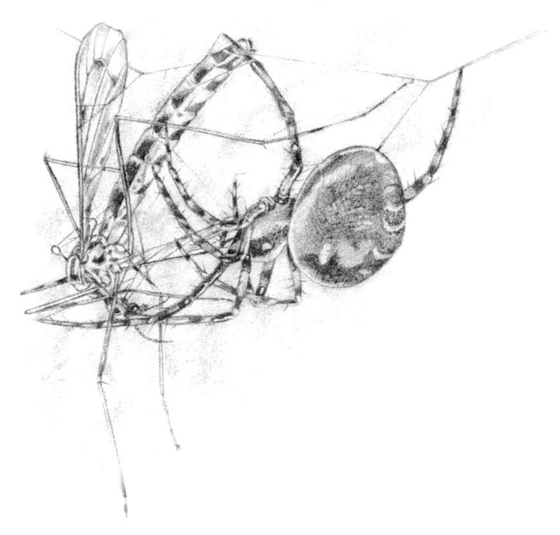

Fig. 4.6 The cave spider, *Meta menardi*, prepares to silk-wrap a cranefly, *Limonia rubeculosa*, which has blundered into her web.

behave like eucladioliths, in that the growing alga directs the accumulating calcite deposition towards the light of the cave entrance. Such light-orientated stalactites must surely occur in the entrances to British and Irish caves, but if so, I know of no mention of them in the literature.

Finally, Jefferson (1989) recorded fully developed specimens of a higher plant – the toothwort, *Lathraea squamaria* – growing and flowering in complete perpetual darkness in Porth-yr-Ogof cave in South Wales. The explanation for this, at first sight, bizarre record is that toothwort, unlike most flowering plants, is not green and does not photosynthesize. It is completely parasitic, deriving its food from tree roots. In this case, the limestone is sufficiently fissured for some tree roots to have breached the roof of the cave, and the toothworts are attached to these, receiving their food in the form of photosynthetic products piped directly down to them from the sunlit canopy above via the tree's vascular system. Unlike the tree roots, the toothworts' flowering shoots cannot penetrate the limestone to reach the surface, and presumably no seed can be set, but vegetatively the plants seem to be thriving.

Invertebrates in cave entrances

Cave entrances are often rich in organic detritus which supports a wealth of invertebrate life. I do not propose to give a detailed account of all the species encountered, as they include a good proportion of the typical fauna of soils and the woodland floor which are not the concern of this volume. However, some invertebrates are so frequently associated with cave entrances that their presence is clearly not accidental, and these we shall now consider.

The place to search for this characteristic entrance fauna is close to the limits of light penetration, on the walls and ceiling of the cave and in its cracks and crevices. The relatively predictable gradients in microclimate within the deep threshold allow its inhabitants to choose to some extent the optimal conditions in which to pursue active lives, or pass periods of quiescence.

Several insect species (two moths, three Diptera and one caddis fly) seek shelter in the deep threshold of our caves in large numbers, year after year, where they can be found occupying the same areas of the cave walls and ceiling for a characteristic period of time before leaving to resume their lives in the outside world. We shall examine their association with caves more closely in the next section of this chapter. On their way in and out, they are trapped by two orb-web spiders (*Meta merianae* and *M. menardi*) which specialize in this niche. Together, these and a few other more cave-dependent species, constitute a characteristic 'wall association' which will also be discussed later.

The large cave spider *Meta menardi* is widespread and common in the deep threshold (and also just within the dark zone) of caves, tunnels, culverts and similar macrocavernous spaces in Europe, Asia (as far east as Korea and Japan), North America and even Madagascar. In Europe its range overlaps with that of *M. bourneti*. Where this happens, the two species exclude each other. In Britain, *M. bourneti* is rare – I know of only one colony in culverts and cellars beneath a large Georgian mansion house in Wiltshire, which it occupies in the absence of *menardi*.

While *Meta menardi* may be regarded as a cavernicole whose distribution encompasses the deeper part of cave entrances, its smaller congener, *M. merianae* is very definitely a cave threshold specialist whose sticky orb webs are almost always sited within the limits of daylight penetration, and in the path of flying insects. *Meta* webs require space and so cannot be built in the smaller diameter cracks in the cave wall and among breakdown blocks. In the deep cave, such cracks are often used by *Porrhomma* spiders, but in the threshold it is more common to find *Nesticus cellulanus* in residence. Its web works on a quite different principle to those of the *Meta* species. It consists of a loose horizontal platform from which threads stretch down and out to the walls and floor of the crack; short sections of these 'anchor' threads, close to their point of attachment to the rock, are studded with droplets of gum designed to trap crawling insects, rather than the flying insects caught in the vertical orb webs of *Meta*. By specializing on different prey in this way, *Nesticus* and *Meta merianae* avoid direct competition for the scarce food resources of the threshold.

Apart from its characteristic array of visiting insects and their predators, cave entrances often contain populations of damp-loving woodlice, snails, millipedes and harvestmen. The most frequently-encountered woodlouse is *Oniscus asellus*, but other ecologically wide-ranging species such as *Armadillium vulgare*, *Porcellio scaber* and *Trichoniscus pusillus* are also common, together with the rather more cavernicolous *Androniscus dentiger*. Typical snails include

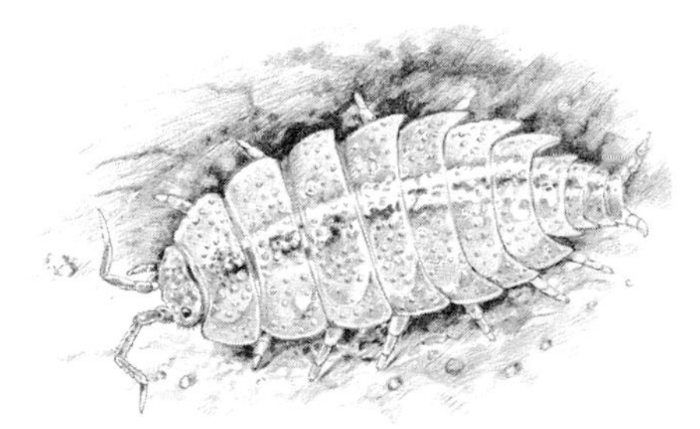

Fig. 4.7 The pretty little pink woodlouse, *Androniscus dentiger*, is a common inhabitant of SUC habitats. It is often encountered in the deep threshold of cave entrances.

Discus rotundatus and various species of *Oxychilus*. The pill-millipede *Glomeris marginata*, though seldom found in the cave proper, is frequently seen in the threshold as are various julid millipedes, and long-legged harvestmen, *Phalangium opilio*, may sometimes collect in large numbers and do a kind of bobbing dance on the cave wall.

Birds

The Swallow (*Hirundo rustica*) has developed such a close reliance on buildings for its nest-sites over most of Western Europe, that it is easy to lose sight of the fact that it must originally have nested in 'natural' sites. Cave-nesting by Swallows was commonplace along the western seaboard of North America well into the 19th century, and was said to be widespread in the Caspian/Kazakhstan area as recently as the 1950s. The habit also persists in some parts of the British Isles to the present day. A century ago, the great Derbyshire ornithologist, Revd. F.C.R. Jourdain, wrote of having found

"a [Swallow's] nest containing eggs, and remains of former nests in a little cave on the side of Bunster, only a few feet from the ground, so that I could look down into it. A pair were also nesting in the roof of Dove Holes on June 19th 1883."

Both caves are in Dovedale, where suitable sites abound, but the increased level of disturbance caused by walkers and climbers seems to have discouraged cave-nesting by modern Swallow populations in recent years.

Elsewhere, Swallows have been seen nesting in a sea cave at Langness on the Isle of Man in 1946, in an open manhole on the side of a Derbyshire colliery tip in 1976, in the concrete shaft of a derelict tin mine in Cornwall from 1976 onwards, in Windy Knoll Cave near Castleton in 1977 and in Giant's Hole in 1986. A substantial colony of Swallows was recorded by V.R. Tucker (*British Birds* 77:210) as regularly nesting inside a disused underground oil depot on the outskirts of Plymouth, Devon, over a 25 year period to 1984; in some cases building against the smooth, vertical, concrete wall "much as House Martins *Delichon urbica* do".

The cave-nesting Swallows of Creswell Crags, in the north-eastern corner of Derbyshire, have been the subject of a recent study by Cliff Davies, who has unravelled a fascinating story of fifteen thousand years of cave-occupation at the site. Creswell Crags is a narrow (100 m wide) gorge running east-west through Magnesian limestone. For about 600 m, the gorge is flanked by 20 m high limestone cliffs containing five major caves, plus many smaller holes, fissures and crevices. The preserved humeri of adult and juvenile Swallows dug from sediments in Pin Hole Cave have been dated back to the late Pleistocene, suggesting that the cave was even then being used as a breeding site. Swallow bones occur in three of the four succeeding strata, with partly ossified juvenile bones present at the youngest levels, dated at *c.*10,000 years b.p. In 1985, six nests were in use, in Pin Hole Cave, Boathouse Cave, Mother Grundy's Parlour, 'C9' cave, Robin Hood's Cave and the cliff face nearby. The following year, perhaps as many as ten pairs were nesting in the vicinity and this pattern has continued to the present. In 1967, the area was designated a Site of Special Scientific Interest for its archaeological remains and steel grilles were fitted over all the cave entrances in 1976/7 to control public access. Davies considers that the protection from disturbance afforded by the grilles has been

Fig. 4.8 Swallows, *Hirundo rustica*, nesting in the caves of Derbyshire's Creswell Crags.

an important factor in sustaining the cave-nesting behaviour of Creswell's Swallows.

Swallow bones have also been found at two more recent prehistoric sites: at Dowel Cave, Earl Sterndale (dated 4000 years b.p.) and at Ossum's Eyrie in the Manifold Valley (dated 2000 years b.p.). At this last site, the bones are part of a collection of other remains which probably accumulated below a raptor's nest. Bones of one or more juvenile Golden Eagles *Aquila chrysaetos* occur in the deposits, strongly suggesting that the cave entrance was used as a nesting site by this species.

Swifts (*Apus apus*) nowadays mainly nest in the roof spaces of houses, which they enter by squeezing through gaps in the eaves. They also occasionally nest in small-diameter cracks and caves in cliffs, and sometimes in tree holes. 'Cave-nesting' swifts select deep, dark clefts which give the maximum protection against predators and adverse weather. Their nest, like that of the Swallow, is a shallow, open cup, placed on a flat surface so that the safety of the eggs and chicks depends on the care with which it is concealed. It would appear that use of mesocávernous cracks represents an archaic behaviour similar to that of Swallows' use of caves.

Dippers (*Cinclus cinclus*) regularly nest in the entrances to caves in the Afon Mellte valley in South Wales and Lathkill Dale in the Peak District; while Red-billed Choughs (*Pyrrhocorax pyrrhocorax*) frequently site their bulky, wool-lined nests in the inaccessible depths of sea caves along the coasts of Pembrokeshire and in Scotland and Ireland.

Other birds which are commonly found nesting in the threshold of caves and smaller holes in cliffs are Rock Doves (*Columba palumbus/livia*), Jackdaws (*Corvus monedula*), and wrens (*Troglodytes troglodytes*).

A place of shelter

Insect visitors in caves

Hibernating insects are presumably attracted to caves by much the same features that attract bats. Principal among these must be the relative predictability of the subterranean environment which ensures that as the animal moves away from an entrance, it has an increasing chance of finding itself in an atmosphere which becomes progressively warmer and less subject to changes in humidity and temperature. Little work has been done on the behavioural ecology of insect visitors in caves, but I hope that the few observations given here may suffice to encourage other naturalists to look deeper.

The Tissue Moth, *Triphosa dubitata*, and the Herald Moth, *Scoliopteryx libatrix*, are so typical of cave entrances that they must be familiar to most reasonably observant cavers. Both are moderately large insects, with a wingspan of around four centimetres. The wings of the Tissue Moth are typically held flat against the cave wall in a half-spread position, so that its outline roughly forms an equilateral triangle. It is patterned in wavy bands of dirty-brown, -green or -pink on a silver-grey background. The Herald at rest folds its wings over its back in the shape of a steep-pitched roof sloping up and widening away from its pointed 'nose', giving it a solid three-dimensional shape. It is a rich russet-brown with white and crimson markings on the wings.

Why, of all the 2000 plus species of Lepidoptera in our islands, just these two should take to caves is far from clear. Their caterpillars feed in the usual way on vegetation outside the cave (the Tissue Moth on buckthorn or blackthorn and the Herald on willow or poplar), but soon after emerging from the chrysalis, the adults fly into caves or man-made substitutes, and settle on the walls or roof. It seems that for part of the time spent underground there is a 'dia-

Fig. 4.9 The tissue moth, *Triphosa dubitata*, overwinters in the deep threshold of caves; changing its position periodically, probably in response to climatic cues.

Fig. 4.10 A group of female golden caddis flies, *Stenophylax permistus*, spend part of the summer clinging to the walls and roof of humid cave passages. It seems that their ovaries require this period of diapause in order to properly mature.

pause' in the ovaries – a period of suspended development – and that this is necessary before the female can produce eggs. In Europe it has been found that adults may spend as much as ten months underground, but no systematic study of the phenomenon has been published in Britain or Ireland.

During their time in caves, Herald Moths become very torpid and can often be seen beaded with moisture on passage walls and ceilings well within the dark zone. Tissue Moths seem to prefer to settle rather closer to the cave entrance and lower down on the cave wall. They are less torpid and may be roused by gently breathing on them, when they may immediately start to 'shiver' their flight muscles (the usual trick used by moths to raise the body temperature to operational flying temperature), causing the wings to vibrate, and may even attempt a short flight to another part of the cave wall. Tissue Moths do not remain in the same place throughout the winter, but seem to move around from time to time. It would be interesting to know whether such migrations follow a pattern related to changes in the cave microclimate, or within the animals themselves, and if so, what the precise physiological purpose may be.

A similar phenomenon occurs with the mosquito, *Culex pipiens*, which overwinters in vast numbers in the threshold of caves or similar sheltered habitats. Fortunately for cavers, this is a species which rarely, if ever, bites humans. Again, it is only inseminated females which go underground. Apparently they require a spell of several months' quiescence before going out in the spring to take a blood meal (usually from a bird) and eventually to lay their eggs in surface waters. Since there are three or four generations a year, this quiescent diapause is clearly not a feature of every succeeding life cycle, but there is some evidence that unless it occurs at intervals of a few generations, the reproductive vigour of the population declines rapidly.

A minor variation on the same theme is provided by the limnephilid caddis fly *Stenophylax permistus*, a large, golden-buff coloured species recorded frequently in caves throughout its range. When resting, the adults hold their their wings together over their backs in a roof-shape and stretch their long antennae straight out in front of the body, giving them an unnaturally stiff appearance. Like the moths and the mosquito, *S. permistus* enters caves only as an adult and seems to require a resting stage in the development of the ovaries. In this case, however, it is part of the summer which is spent underground.

Leiodid beetles of the genus *Choleva* are in a sense pre-adapted to cavernicolous life, since some are associated with ants' nests, and many feed on detritus and fungi. Adults of *C. agilis, spadicea, glauca* and *oblonga*, are occasionally found in the cave threshold, concealed inside clay cells which they dig out and construct using their mandibles. As with *Stenophylax permistus*, this behaviour seems to be connected with the need for a period of summer diapause. The beetles begin building towards the end of spring, closing themselves in completely for the duration of the summer months. Although the onset of this behaviour may be triggered by day length (or some other such seasonal cue), it is not clear what cues the re-emergence of the beetles in the autumn, but it is possible that aestivation is timed by an internal metabolic 'clock' which wakes the animals when their time is up.

Two other insect species which commonly visit the cave threshold are the pretty little cranefly, *Limonia nubeculosa*, and the heleomyzid dung-fly *Heleomyza serrata*. Not a great deal is known about the reasons for the presence of either species in caves. Neither is thought to feed while underground, and indeed it is thought that very few crane-flies feed anyway in the adult stage. *Limonia* is a summer visitor and is known to tolerate a wide range of cave temperatures and humidities throughout its European range (from 10–20+°C, and 50–100% relative humidity). *Heleomyza*, on the other hand, arrives in caves in the autumn and may remain in a semi-torpid state right through the

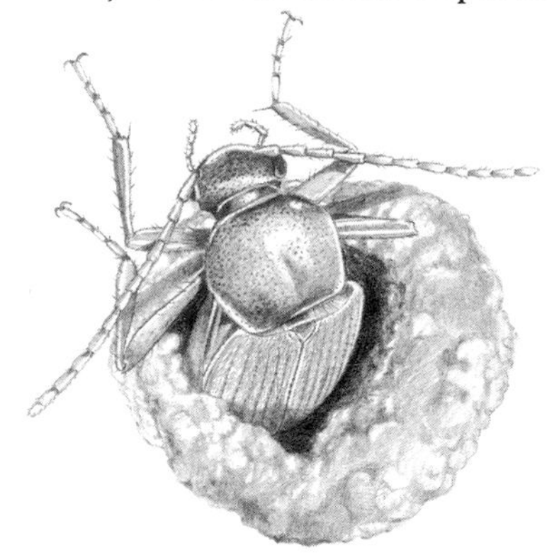

Fig. 4.11 At the end of its summer diapause, a *Choleva* beetle breaks out of the clay 'igloo' in which it had sealed itself the previous spring.

following year and indeed until it eventually dies from starvation or fungal attack. This rather dozy behaviour is difficult to explain, but seems to be a particular feature of populations which settle in completely dark areas of the caves where the humidity, in summer at least, is close to saturation. I can only guess that the flies may be the victims of seasonally-reversing air currents – a common feature in the large Welsh systems where I have observed them most often. I imagine that the flies may enter the cave by following a cool inward-flowing winter airstream, moving gradually further from the entrance until they find themselves in a spot whose temperature and humidity is to their liking, and where they settle to see out the winter. Come spring, the draught in the cave slows, stops, and eventually reverses. As it does so, the relative humidity of the cave air increases dramatically towards saturation, and the flies begin to experience water poisoning. This first debilitates, and later kills them, or renders them easy prey to insect-eating fungi.

Bats

Throughout the world, around 1000 species of bats are known, many of these being cave-roosting. Of the two sub-orders, the Megachiroptera, or fruit bats, are confined to warm climates and have no European representatives, while the insect-eating Microchiroptera range from northern Norway and Alaska to Cape Horn. There are sixteen Microchiropteran families, of which only two, the Rhinolophidae (Horseshoe Bats) and Vespertilionidae (Evening or Vesper Bats) occur in Britain and Ireland. Both of our Horseshoe Bats and six out of our 13 species of Vesper Bats are known to shelter regularly in caves, while a further three Vespertilionids make use of them occasionally. With the rise of civilization in our islands, most of these 'cave bats' have learned to make do with man-made substitutes when the real thing is not readily available; but the two Horseshoe Bats are still very much dependent on natural caves over a good deal of their respective ranges.

During the course of a year, an individual bat may regularly shift between a large number of different roosting sites. Over the last three decades, naturalists have begun to uncover a pattern in these seemingly haphazard movements, and to appreciate the extent to which each roost is carefully selected to meet seasonally varying requirements.

The picture which has emerged can best be illustrated by considering our most cave-dependent bat, the Greater Horseshoe, *Rhinolophus ferrum-equinum*, a species which is becoming increasingly rare over much of its range in South-west England and South Wales and whose association with caves is therefore of particular interest to the naturalist. From the moment it achieves the skill to fly, the young horseshoe bat begins to make a series of choices of where to live and one of the most important criteria which it employs is temperature. Unlike most larger mammals, bats do not maintain their body temperature within narrow limits, but instead allow it to vary according to requirements – the normal body temperature range for a Greater Horseshoe Bat being between 5° and 40°C.

During July, when the mothers are nursing young and need to maintain their body temperature close to the higher figure, the ideal ambient temperature in the roost is about 30°C. Until recently many Greater Horseshoe nurseries were sited in caves, where the ambient temperature would normally be 9–12°C. The bats circumvented this potential problem by selecting chambers with a domed roof which would trap the warm air produced by their densely-packed bodies. By regulating their numbers and their spacing on the cave ceiling, the nursing mothers were able to maintain an optimal temperature for their young.

Towards autumn, the nursery clusters begin to break up, with the adults heading for cooler areas while the young bats generally stay together, often in the nursery site, until October. With the thinning out of the colony, the temperature in the main roost gradually falls, and the juveniles drift into torpor for a part of each day, rather than remaining alert and awake. In this way they conserve some of the energy which they would otherwise expend in keeping warm. As the weather worsens, and food becomes scarce, they gradually spend an increasing proportion of each day in a state of torpor, with their body temperature down to around 12°C. By November, most bats are torpid for days at a time, waking only during warm weather when feeding is possible.

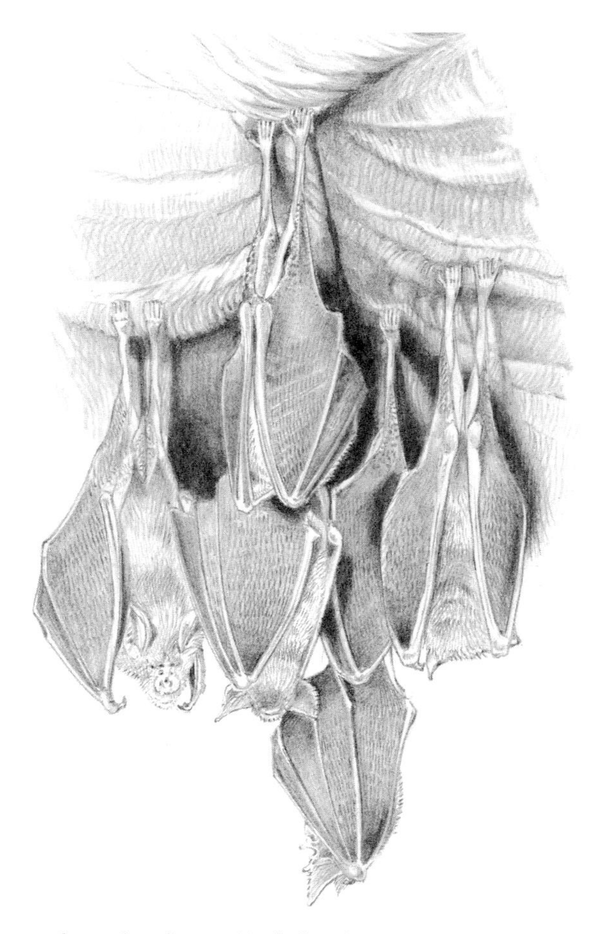

Fig. 4.12 Greater horseshoe bats, *Rhinolophus ferrum-equinum* are our most spectacular cave-roosting species. A hibernating group like this is likedly to consist of juveniles – adults, both male and female, prefer to roost singly.

Bats usually reach their maximum body weight around October of each year, before the onset of hibernation, and although they continue to feed whenever they can throughout the winter, the general trend is for them steadily to use up their fat reserves until the following May, at which time they may have lost a third of their mass of the previous autumn.

As winter progresses, Greater Horseshoe Bats arouse about every 10–15 days in order to regulate the composition of their body fluids by drinking and urinating. Such arousals may simply involve a rise in metabolic rate to bring the body temperature up to about 35°C, or, if the ambient temperature at the roost is found to be unsuitable, the roused bat may fly off in search of a new roosting site. This may be in the same cave, or anywhere up to 50 km away. The sites are precisely chosen according to the bat's body mass. Heavy bats choose warm roosts, while thin bats are found in cool places. Experienced adult bats are better at selecting appropriate roosts – they tend to occupy the same sites at about the same period each year and maintain their body weight within narrower limits of variation than young inexperienced bats.

As the bat's food reserves dwindle and its mass falls, gradually cooler roosts are chosen, so that by March, a bat which began the winter in a roost at 12°C, will be found at about 6°C. In April, towards the end of the hibernation period, the lengths of torpor shorten and the bat begins to emerge to feed on warm evenings. Cool roosts are chosen close to good feeding areas so that after digestion (which may take up to two hours), the bat can quickly drop its temperature, using its wings as radiators, and hence conserve energy.

For reasons which are not fully understood, juvenile Greater Horseshoe Bats, with some immatures and a few adults, often form roosting clusters, while old adults, both male and female, mostly roost singly. A few males behave in a territorial manner, especially in the autumn when mating occurs.

Other bat species have their own different roosting requirements and preferred range of temperatures. While Greater Horseshoes select the warmest roosts, the tiny Lesser Horseshoe Bats, *Rhinolophus hipposideros*, prefer slightly cooler sites and can often be seen sharing the same winter roosts as their larger cousins, but in more exposed parts of the cave ceiling.

Lesser Horseshoes have a wider geographical range than Greater Horseshoes, being found in Western Ireland and North Wales, as well as South-west England and South Wales.

The Barbastelle, *Barbastella barbastellus*, and Brown Long-eared Bats, *Plecotus auritus*, have the coolest requirements of all, choosing winter sites close to the cave entrance or crevices in cliffs, where they can be found with a body temperature as low as 0°C. The former has been recorded in England as far north as Yorkshire, but is rarely seen, while the latter is widespread throughout Britain and Ireland. When asleep, a Long-eared Bat folds the pinna of its enormous ears backwards under the wings, while the tragus (a vertical skin flap which guards the ear orifice) continues to project upwards, looking rather like ears in themselves, so that the bat is not immediately recognizeable.

The remaining six species which use caves all belong to the genus *Myotis*. Four of these, the Whiskered (*M. mystacinus*), Brandt's (*M. brandti*), Natterer's (*M. nattereri*) and Daubenton's or Water Bat (*M. daubentoni*) are widely-distributed in Britain and probably in Ireland. While of the remaining species, Bechstein's Bat (*M. bechsteini*) is rare and confined to southern England, and while the Mouse-eared Bat (*M. myotis*) is now reckoned to be extinct on this

Fig. 4.13 During hibernation, Brown Long-eared Bats, *Plecotus auritus*, tuck the pinna of their ears behind their wings, leaving the long tragus pointing forwards, making them difficult to recognize.

side of the English Channel. The Myotis Bats may be found in winter roosts with temperatures in the range 0–9°C, with most preferring a range of 2–5°C.

While Horseshoe Bats hang by their hind feet from projections on roofs and walls of the roost, their wings tightly wrapped around the body like a cloak, most other cave bats choose narrower crevices of mesocavernous dimensions into which they crawl for up to several metres, and Daubenton's Bats also wriggle into loose scree or boulder piles in caves. Dr Bob Stebbings, a foremost bat expert, reports finding one such bat after hearing squeaks of protest as he walked over a pile of rocks in a Devon cave. It is possible that bats seek out such enclosed spaces for the same reasons as do cave invertebrates – to escape the drying effects of winter airflows, but this is not known for certain.

Because cave-roosting bats may not always be easily visible to the human visitor, it is difficult to estimate empirically which caves are most important from the point of view of bat conservation. However, with our improving knowledge of the biology of bats, it is possible to assess the likely value of particular caves as roosting sites by measuring temperatures and rates of air flow, by observing the size and configuration of cave passages, and by taking into account the geographical location, altitude and proximity of the cave to other known roosts and feeding areas. The best caves in terms of numbers and diversity of bats using them are those in southern lowland areas bordering marshes and woodlands, which have varied chambers, passages and crevices with a dynamic environment.

I mentioned earlier that many cave bats have adapted readily to man-made 'pseudo-caves', such as railway and canal tunnels, cellars, fortifications, ice houses, the spaces in the walls of houses (particularly those with modern 'cavity-wall' construction) and mines – above all the stone mines in the limestones of Gloucestershire, Wiltshire and Avon. Some of the latter were started during the Roman occupation and now form a last stronghold for the region's dwindling populations of horseshoe bats. Modern bat enthusiasts from English Nature and the Flora and Fauna Preservation Society's network of 'local bat groups', taking a cue from from the Romans' (unwitting) efforts in practical bat conservation, have taken to opening up blocked tunnels and have even dug a special purpose-designed bat cave' in chalk at Whipsnade. We will return to the topic of bat conservation in a later chapter.

Bat parasites

Bats in caves may harbour a range of ectoparasites, such as the curious wingless flies of the family Nycteribiidae which have a flattened, apparently 'headless' body, long legs armed with hooked feet for clinging on to their hosts, and a 'pupiparous' life cycle in which the larval stages are completed within the body of the female fly. Two species of nycteribiids have been reported from bats roosting in caves: *Nycteribia (Stylidia) biarticulata*, which occurs on the Lesser Horseshoe Bat, and *N. (Nycteribia) kolenatii* which lives on Daubenton's Bat.

Fleas are rather common on bats, and two species, *Ischnopsyllus simplex* and *I. hexactenus* have been collected from cave-roosting Natterer's and Brown Long-eared Bats respectively.

Most bats harbour ticks, much the most common of which is the ixodid Bat Tick *Ixodes vespertilionis*, whose geographical distribution extends throughout Europe, Asia, Africa, and even Australia. It is very much a cave-associated species, considered by Vandel (1965) to be essentially troglobitic. Only the female tick sucks blood, the male, which may be very common on the walls of bat caves, seems to have no need of food. Young ticks gain access to bats from the walls of the cave, and usually attach themselves to the back of the neck or under the jaw, so that the bat cannot readily remove them. When full of blood, they drop off and can be collected from the cave floor. Another tick, the argasid *Argas vespertilionis* has been found in roosts of the Greater Hoseshoe Bat in Devon.

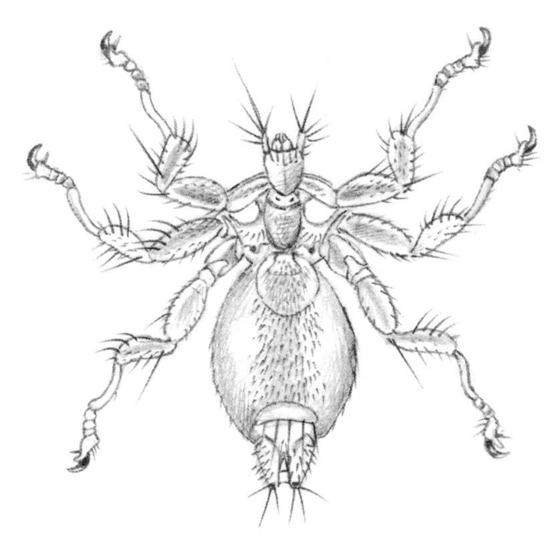

Fig. 4.14 The parasitic bat-fly *Nycteribia kolenatii* which lives on Daubenton's Bat.

Nursery roosts of Greater Horseshoe Bats may also become infested with mesostigmatid mites such as *Spinturnix euryalis*, which apparently attack only the young bats. As the juveniles get older, the infestation disappears. Another parasitic spinturnicid mite, *Periglischurus rhinolophinus*, together with the macronyssid *Hirstesia sternalis*, have been collected from Lesser Horseshoe Bats in Devon caves.

Other mammals

In Upper Pleistocene times, many caves in the British Isles were used as permanent dens over long periods of time by generations of Spotted Hyaenas (*Crocuta crocuta*) and Brown Bears (*Ursus arctos*). Other carnivores, notably Cave Lions (*Panthera spelaea*), Wolves (*Canis lupus*), Lynx (*Felis lynx*), Wolverine (*Gulo gulo*) and Red Fox (*Vulpes vulpes*) also appear to have frequently entered caves, and the remains of all of these animals and of their prey are widespread and common in fossil bone deposits (see Chapter 6). At the present time, only two mammals, other than bats, appear to make regular use of caves. They are the Badger (*Meles meles*) and the Brown Rat (*Rattus norvegicus*).

William Boyd Dawkins, writing in 1874, described a subterranean encounter with Badgers during an early visit to Goatchurch Cavern on Mendip.

"The cave is the resort of numerous badgers. On hiding ourselves in one of the transverse fissures, and throwing our light across the horizontal passage, these animals ran to and fro across the lighted field with extraordinary swiftness, and had it not been for the white streaks on the sides of their heads, which flashed back the light, they would not have been observed."

Alas, Badgers have long-since deserted Goatchurch owing to human disturbance – this being perhaps the most heavily visited cave in the British Isles. However, in Cheddar Gorge, the appropriately-named Brock Hole has retained its longstanding Badger colony. Naturalist Michael Woods reports that the only human-sized entrance to the cave is rather difficult to reach, being approached by a long and tortuous traverse among trees on a steep and slippery slope. He doubts that this is the entrance favoured by the Badgers because they would be unlikely to be able to carry bedding to it, and also because it is filled with an unpleasant odour from dung pits on the cave floor; and Badgers are not in the habit of fouling their own front doorstep. A more suitable Badger-sized entrance is found further back on the slope. It is a narrow excavated cleft in the rocks – too small for humans to enter, but just the kind of naturally-fortified portal which would suit a group of Badgers. Four or five of the animals have been seen on the track leading to the cleft, which at one time held some bones of a Fox.

In the same part of Cheddar Gorge, another rocky cleft appears to be used by Badgers as a daytime 'lying up' den. Inside is a metre-high stack of grass, which constitutes the day nest.

Badgers have been reported at least 100 m (and perhaps as much as 200 m) into Glencurran Cave, which opens into the ancient Hazel woodland of Glen Curran in Co. Clare. The entrance passage is four metres wide, by three metres high and contains a number of pits about 15–20 cm deep and a metre across, dug into the sediment floor some 15–30 m inside the cave and containing vegetation and fur. Deeper in, the cave becomes blocked with sediment almost to the roof. Dave Drew and his Dublin-based cavers spent several years digging out sand in an effort to widen the restricted roof-tube into a human-sized passage. For the whole length of their dig, they found the route lined by the footprints of badgers and made several sightings of the animals themselves. The footprints were eventually lost when coarser sediments were reached, but it is quite possible that the badgers utilize the still unentered depths of the system beyond the current limits reached by cavers.

By the River Wear, near Westgate, in Durham's only area of cavernous limestones, Slit Woods offers a rare haven for Badgers. The wood is traversed by the Middle Hope Burn, flanked by a 10 metre-high cliff. Upstream of the cliff, by the side of the Burn, sits a hole which looks as if it was formed by the removal of a single peg of limestone, 50 centimetres square. This marks the entrance to the sett, safely encased in solid rock beyond the reach of diggers and their dogs. High up in the cliff, a narrow fissure marks an accidental back door to the sett – dug out from within and quite unusable in its precarious position.

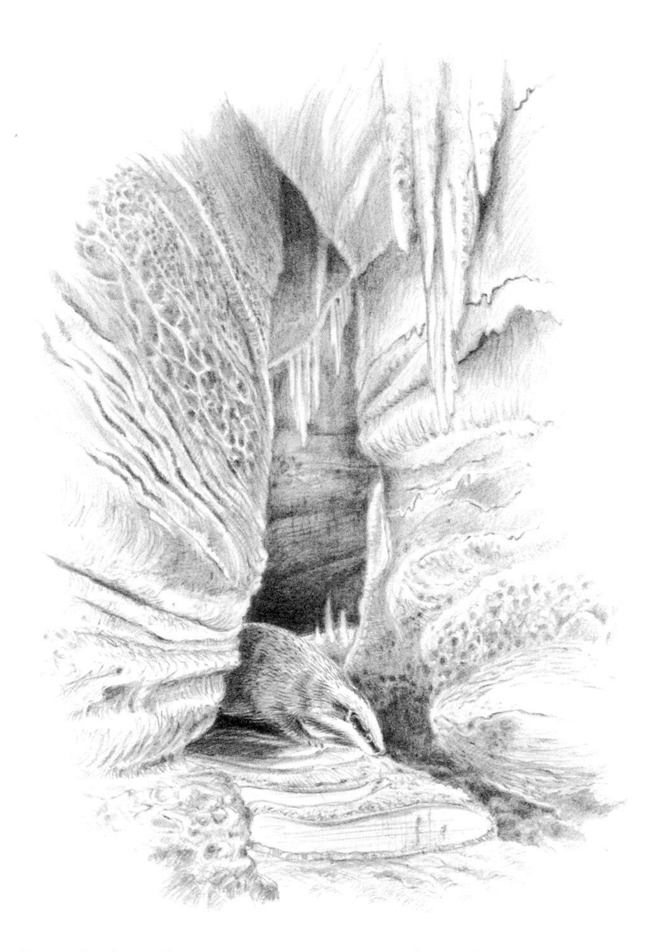

Fig. 4.15 Badgers, *Meles meles*, may use caves as ready-made setts.

Badger setts are often dug beneath tree roots or large rocks which provide security against accidental collapse or the unwelcome attentions of man. Limestone caves, if naturally dry and filled with easily dug sediments, can provide an ideal home for Badgers, which in any case favour the earthworm-rich, well-drained soils associated with karst.

Brown Rats, by contrast, do little or no digging in caves. They presumably go below ground in search of food, and in this they are encouraged by the actions of a few irresponsible farmers and slaughterhouse owners who use cave entrances as convenient disposal chutes for unwanted offal and other organic rubbish. One Mendip caver reported that, on his way out of the malodorous Stoke Lane Slocker cave on Mendip with a failing carbide lamp, he lost his way in the entrance series and strayed up a damp crawling-sized passage which gradually lowered to an impassable bedding slot. Sensing strange movements in the shadows ahead, he stopped to fettle his flame, and having done so, found himself face to face with "a living wall of rats", which "seemed more curious than afraid". The cave has a reputation for harbouring the causative agent of Weil's Disease, a particularly nasty pathogen known to be carried by *Rattus norvegicus*.

Waifs and strays

Fishes

World-wide, about 60 species of blind, white cave fishes have been found mainly in tropical and subtropical countries. Curiously, none at all are recorded from Europe, although several highly specialized forms occur in central USA just south of the furthest limit reached by the Pleistocene glaciations. The strangest cave fish are those found in the groundwaters which underlie arid regions of Somalia, India and Texas. They are entirely without eyes and pigment and have extremely skinny bodies, a feature taken to the limits by the symbranchids *Pluto infernalis* from Brazil and *Anommatophasma candidum* from Australia, which are something like 50 times as long as wide.

In Britain, some ten species of freshwater fishes have been recorded in caves. To date, none of these has been shown to maintain cavernicolous populations, and none possesses any signs of genetic modifications which might equip it for a cavernicolous existence.

The commonest fish by far in British caves is the Brown Trout, *Salmo trutta*, which has been reported from more than fifty caves in Yorkshire, Durham, South Wales, Derbyshire, Fermanagh and Clare – though, strangely, there appear to be no recent records from Mendip. The first traceable reference to trout in caves is in an unpublished manuscript prepared by John Strachey in about 1730. In referring to Wookey Hole, he writes:

"This cave gives harbour to very fine trouts which on every flood stock the River without."

John Hutton's famous *Tour* of 1780 records that:

"Trouts of a very protuberant size have been drawn out of the Mere [of Meregill Hole in Yorkshire], where they have long been nourished in safety, their habitation being seldom disturbed by the insidious fisherman."

Trout probably enter caves by accident – they are known to move downstream, particularly at night and at certain ages and times of year, and if the stream enters a cave, so will the fish. The other possibility, that the fish deliberately enter caves during upstream migrations and remain there out of preference would seem less likely as their whole lifestyle is quite vision-dependent. Moreover, cave specimens on the whole tend to be less well-fed than their surface counterparts, suggesting that selection pressures would tend to favour a routine avoidance of caves among members of populations inhabiting streams in karst areas.

The frequent reports of "white fishes" in subterranean streams and lakes are almost always attributable to *S. trutta*, although to date none of the specimens collected has proved to be fully depigmented. In the darkness of the cave, trout which are merely pallid may seem pure white, but a close examination shows that the body is peppered with small coloured or black spots which mark the position of the pigment cells, and there is usually a distinctive row of orange-pink spots along the lateral line, and touches of pink on the fins and back. Some bottom-dwelling fishes such as Plaice and Soles can manipulate their various pigment cells, or chromatophores, with such dexterity that they can almost instantly take on the pattern and hue of their surroundings and so escape detection by predators. Trout can also change the appearance of some of their chromatophores (but not those containing the orange-pink carotenoid pigments) and so alter their basic colour – though this behaviour has little or nothing to do with camouflage. The dark markings of fishes are produced by the pigment melanin, contained within star-shaped melanophores which are distributed in a characteristic pattern over the body surface beneath

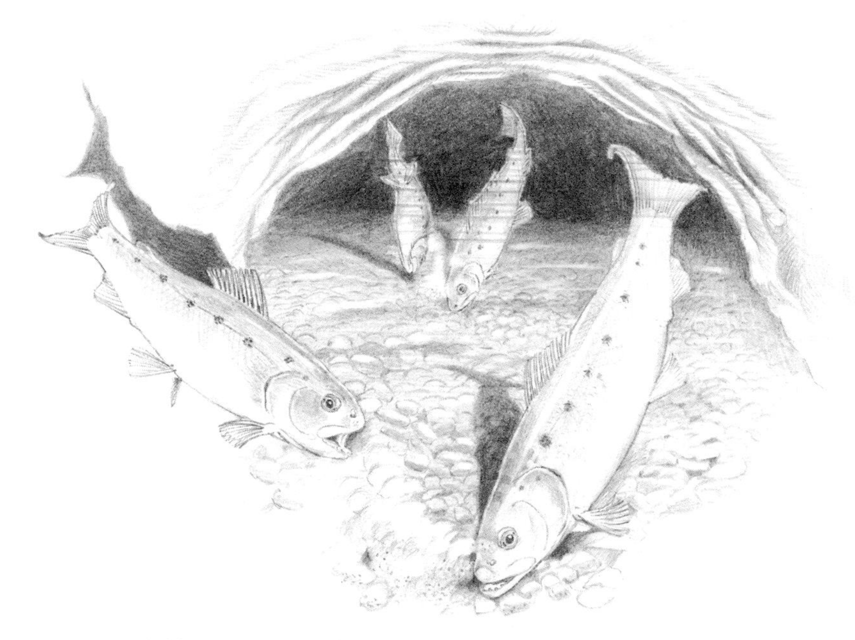

Fig. 4.16 Pallid brown trout, *Salmo trutta*, grub for food in the gravel of Lake Avernus, in Yorkshire's Ingleborough Cave.

the scales. 'Blanching' is achieved by withdrawing pigment from the radiating branches of the melanophores into the centre of each cell, exposing the underlying iridiophores which are filled with reflective guanine. The pale colour of cave trout is therefore no more than a temporary effect, and when removed from their cave and placed in a lighted tank the 'white fishes' are found to regain their normal colouration within 24 hours.

Contrary to popular belief, cave trout are not blind, although obviously while underground they must rely on other senses such as touch and smell. When blinded, salmonid fishes invariably turn black, rather than pale, and the few cave trout which have been examined under controlled conditions have been found to have structurally normal eyes and at least some degree of vision. It is therefore difficult to explain the consistent reports by numerous cavers and cave divers that trout encountered underground generally show no reaction to lights shone on them. A typical account is given by Dick Glover in the *Journal of the Craven Pothole Club* (1978):

"Whilst exploring the small oxbow near the head of Lake Avernus [Ingleborough Cave], I noticed a small, white fish left stranded in a small pool by the low level of the stream itself, which apparently was very interested in eating the dubbin on my boots. This fish, about 3 inches [8 cm] long, appeared to be completely insensitive to the light from my headlamp, but reacted immediately to any disturbance of the water. After several attempts I managed to grab it and examine it closely. It appeared to be identical in shape, size, etc. to the common Brown Trout, often found in pools in mountain streams. However, apart from a few spots ... it appeared completely devoid of normal pigmentation. The pupils of its eyes were, however, dark in colour and the eyes themselves appeared to have developed normally ... Swimming down Lake Pluto [further into the cave], I noticed a fairly large population of similar fish resting on the bottom and on underwater ledges. I estimate I saw at least two dozen fish ranging in length from 3 inches [8 cm], to upwards of a foot [30 cm] ... None of the fish appeared to be disturbed by the illumination provided by my headlamp and diving torch, but all reacted swiftly to any pressure vibrations caused by splashing or sudden movements of my arms and legs underwater."

Brown Trout are essentially carnivorous, preferring to feed on aquatic insects (nymphs, larvae and adults), Crustacea and molluscs, although large individuals will also take smaller fishes, including other trout. Their normal feeding technique is to watch from a position in mid-water for any movement by potential food organisms which are then captured in a rapid swoop – a style reminiscent of the kestrel by a motorway verge. In caves, where it is totally dark, the variety of prey and the ease of capture will both be greatly reduced, yet trout seem able to feed underground, a fact confirmed by examination of the stomach contents of specimens collected from Lake Avernus at the Ingleborough Cave end of the Gaping Gill system in Yorkshire. The lake is separated from Fell Beck (considered to be the most likely point of entry into the underworld) by three kilometres of cave tunnels and a 110 metre-waterfall and trout entering by this route must take some considerable time to arrive at their destination. As food is rapidly passed from the stomach to the intestine in trout, it is inconceivable that the stomach contents of Lake Avernus fishes could represent the remains of food taken outside the cave and it must be concluded that the trout are feeding in the cave itself. Lake Avernus contains a substantial population of the stream shrimp, *Gammarus pulex*, which would seem to represent an adequate source of nutrition for the fishes, although how they manage to catch them is something of a mystery. A clue is provided by the anatomy of a cave trout from Ingleborough Cave which had a worn-down lower jaw, giving its snout a blunt appearance and suggesting that it had spent a great deal of its life grubbing for food in the gravel of the cave floor. Examination of its stomach contents revealed, in addition to the expected amphipod remains and detritus, a number of small stones – suggesting that a certain degree of hit-and-miss is involved in this feeding technique.

The one unequivocal report of *Salmo trutta* feeding in a cave comes from Frearson (1964), who recounts observations of a tame trout called Dick, which lived inside Bruntscar Cave in Yorkshire, and would come to be fed when called. Frearson records that:

"The trout took any kind of food – fish, flesh or fowl. It enjoyed beef and mutton best."

Fishes' scales, like tree-trunks, grow concentrically outwards throughout their lifetime and can give information about the age and rate of growth of the individual fish from which they are taken. Comparisons of age and size between scales from trout taken in surface rivers, lakes and caves shows that they grow very slowly underground. Longevity in trout is inversely correlated with growth rate, so the large trout reported by cave diver John Cordingley, Dick Glover and others from the lakes and inundated passages of the Gaping Gill system may be very old indeed, and it would be interesting if one or two could be netted and a scale or two returned for analysis.

Though less conspicuous than 'white' trout, the little, bottom-dwelling Bullhead or Miller's Thumb, *Cottus gobio*, is also regularly reported from caves in Wales and Yorkshire. Bullheads, as their name implies, have a broad, heavy head, armed with backward-pointing spines, and may grow to about 10 cm long. They live in fast-flowing streams, using their enormous pectoral fins to creep about in the sheltered water beneath stones on the stream bed in search of the insect larvae and crustacea on which they feed. The most likely

Fig. 4.17 A bullhead, *Cottus gobio*.

explanation for their presence underground is that they are swept downstream into caves during floods. Underground, they retain their normal greeny-brown mottled colouration. Graham Proudlove reports that one specimen removed from Kingsdale Master Cave in Yorkshire became lighter in colour when placed in an illuminated dish, suggesting that the species may darken its colour in the cave – the opposite response to that of the Brown Trout.

The third fish commonly seen in caves, particularly in Ireland, is the Eel, *Anguilla anguilla*. Adult eels migrate to the Sargasso Sea to spawn, and the young eels then complete the return journey into fresh water, moving up the rivers in large numbers each spring. It is easy to explain their entry into caves, either via resurgences in river beds, or via cave entrances after cross-land travel.

To my knowledge, only seven other species of fishes have been reported from caves in Britain and Ireland. Although they are very much stray observations, I mention them here in the hope that my cave-frequenting readers may be stimulated to contribute records of other species, which must surely enter caves from time to time.

A solitary goby, *Pomatoschistus* sp., probably the estuarine *P. microps*, has been found living in a pool in a section of Otter Hole which lies below the high tide mark near the bank of the River Wye near Chepstow. The pool contained brackish water and there is little doubt that the fish had been accidentally stranded by a tidal surge from the river outside.

Salmon, *Salmo salar*, have been reported in two caves: the Coolagh River Cave in Co. Clare, and Hurtle Pot in Yorkshire. The former record was originally explained in a similar manner to that of *Pomatoschistus* in Otter Hole – the subterranean Coolagh River resurges on the sea shore, and had flooded shortly before the fish was captured. However the specimen in question was a parr (one of the juvenile stages of the Salmon) which can live only in freshwater, so it seems more probable that the fish had entered the cave passively by downstream drift. Presumably, for it to have reached its eventual resting place, Salmon must have spawned in the headwaters of the Coolagh River and it would be interesting to know whether the cave forms a regular Salmon run.

The record from Hurtle Pot is also of a parr, but it is unthinkable that spawning Salmon could have reached the headwaters of Chapel Beck whose waters feed the cave, as the journey would require negotiating a number of large, overhanging waterfalls, including one of six metres. The best explanation I can offer is that Salmon eggs or young fry were transferred in some way to the upper reaches of Chapel-le-Dale and then washed downstream into the cave.

Rainbow Trout, *Salmo gairdneri*, have been reported in Arch Cave, Co. Fermanagh, and Russet Well in Derbyshire. Both are areas where this introduced species is known to occur in rivers, and it seems probable that they will have gone underground in much the same way as Brown Trout do elsewhere.

The Stickleback, *Gasterosteus aculeatus*, has been collected on a single occasion each from Poulnagollum in Co. Clare and from Boho Cave in Co. Fermanagh. Minnows, *Phoxinus phoxinus*, are widely distributed in stony streams throughout our major caving areas, yet there has been only one unconfirmed sighting of a single 'white' specimen in Little Neath River Cave in Wales.

The Pike, *Esox lucius*, lives mainly in large rivers and lakes and would not be expected to be found in caves, so the single unconfirmed record from the Peak Cavern resurgence in Derbyshire must be treated with scepticism, particularly as the river below the cave would seem quite unsuitable for this species.

Finally, cave diver Martyn Farr has reported seeing a white flatfish some 5 inches (12.5 cm) long in the polluted Merthyr Mawr Cave in South Wales. Of the 20 or so species of flatfishes living in British offshore waters, only one, the Flounder *Platichthys flesus* regularly enters fresh water. It is most likely that the fish entered the cave accidentally during a tidal surge.

Unusual impostors

Almost any organism which lives in a karst area may be accidentally introduced into caves, either by tumbling into a hole, or by being washed in by a sinking stream. A really dramatic rescue involving a pet dog or a farm animal will often make headlines in the local newspaper, while the more biologically remarkable cases escape public notice. A swift scan of the log of the Cave Rescue Organization shows that the most likely species to require assistance in caves are sheep (about five per year – most of these, for some unfathomable reason, from the 24 metre-deep shaft of Fluted Hole on Ingleborough), followed by humans (marginally more intelligent), with dogs and cattle lagging far behind. Most cavers will have their own favourite anecdotes – I offer the reader a few of mine.

On January 16th 1797, the *Bristol Mercury & Universal Advertiser* ran the following story:

"Yesterday fe'nnight as two young men were pursuing a rabbit in Burrington Coombe, they observed it take shelter in a small crevice of the rock. Desirous of obtaining the little animal, they with a pickaxe enlarged the aperture, and in a few minutes were surprised with the appearance of a subterraneous passage leading to a large and lofty cavern the roof and sides of which are most curiously fretted and embossed with whimsical concreted forms. On the left side of the caverns are a number of human skeletons lying promiscuously, almost converted into stone."

The rabbit hunters had stumbled on Aveline's Hole, a major archaeological site.

A contemporary version, which illustrates how attitudes to wildlife have changed in the interim, concerns a young rabbit found wet and shivering 400 m down the stream passage of St. Catherine's One – part of the Doolin Cave system of Co. Clare – by members of the University of Bristol Spelaeological Society. The poor, frightened creature was rescued by caver Barbara Sreeves, who, deciding that it needed warmth and comfort, unzipped her wetsuit jacket and tucked the little rabbit in, with just its nose and ears peeking out. Having quietly endured the rest of the caving trip in this position, the rabbit was released out on the surface, where it stood, pensively cleaning its face for a moment or two, before hopping purposefully back down into the cave.

Otter Hole, with its tidal connection into the River Wye is nicely set up to produce the occasional biological surprise. During the early period of exploration of the cave in 1978, two cavers, John Elliott and Jim Hay of the Royal Forest of Dean Caving Club were traversing a ponded-up section just upstream of the tidal sump, when they noticed that the water surface appeared to be 'boiling' with thousands of translucent shrimps. Peering closely into the deep green water, they saw a huge shoal of what seemed to be slender-bodied marine prawns. These turned out to be the brackish-water mysidacean *Praunus flexuosus*, previously unrecorded from caves. The water at this point was only slightly brackish, so it seems likely that the shoal was on its way upstream from the tidal sump when spotted by the cavers. It would be interesting to know how often such a phenomenon occurs, and how far upstream these creatures will venture.

Britain has no native cavernicolous amphibians, but the warmer caves of Slovenia are home to a singular cave salamander, the Olm, *Proteus anguinus*, which is entirely without eyes or pigment. Olms retain larval gills throughout their adult life and are thus able to colonize the flooded tunnels of the phreas as well as vadose cave streams. For a time during the 1930s, a number of these creatures were kept in a tank in the Zoology Department at Bristol University. Eventually, in 1940, it was decided that they had outlived their welcome, and two students in the Department, Brown and Goddard, were dispatched to the Mendip Hills with instructions to release them into a suitable cave. They chose Read's Cavern, on the northern slope of Blackdown, close to Burrington Combe. The stream which enters Read's (called Hunter's Brook) can only be followed for a short distance underground, but has been dye-traced to both the Rickford and Langford risings, which also receive the waters from the East and West Twin Swallets and Ellick Farm Sink. It is clear from the pattern of flow that between all of these sinks and risings lies a lot of undiscovered cave passages. The question is whether the Olms survived, and survive still somewhere beneath the Mendip Hills.

A number of cavers have reported seeing frogs or newts long distances into caves, and there is some evidence that common frogs, *Rana temporaria*, can survive for a considerable time underground – although, faced with an inadequate food supply, they become progressively thinner and eventually die of starvation. The Coolagh River Cave in Co. Clare seems to collect unusually large numbers of frogs, which sit appealingly along the banks of the streamway, prompting a visiting English caver, Dick Willis, to set up his impromptu 'Irish Frog Cave Rescue Association' in the early 1970s. Members devised a technique of inserting a dazed amphibian between the top of the head and the roof of the helmet, where it would remain safe, if somewhat hot, until an exit was made from the cave. Many a passing farmer will have been amazed to witness the cascade of escaping amphibians as cavers stripped off their gear under the gaze of Ballynalackan Castle. Apparently the 'Cerberus Frog Rescue Service' performs a similar function on Mendip, running regular mercy missions into the notorious frog-gulping St Cuthbert's Swallet.

Denizens of darkness

We have moved through the cool twilight of the cave threshold, tiptoeing quietly past the roosts of slumbering bats, and are now entering the perpetually dark domain of the cavernicoles – full-time residents who are able to survive from one generation to another on resources present in the cave. In Britain and Ireland there are no vertebrate cavernicoles, but most of the common groups of invertebrates are represented, as are a host of even simpler organisms: protists, bacteria and fungi. Without going into too much detail, I shall attempt a brief review of these 'denizens of darkness' by Kingdom, Phylum, or Class, as appropriate.

Bacteria

Perhaps the best known and most conspicuous manifestation of bacterial action in caves is the phenomenon of 'moonmilk' – a soft, cheesy calcareous deposit which is widely distributed in temperate and tropical caves. It consists

Fig. 4.18 The "well-shrimp" *Niphargus fontanus*, is a true "denizen of darkness" – a translucent, eyeless troglobite forever committed to subterranean life.

of a complex of all sorts of unusual magnesium and calcium carbonates, some of which are alleged to be associated with particular bacteria. In Britain, quite a range of organisms has been isolated from moonmilk. One of these, the bacterium *Macromonas* sp., is capable of depositing crystalline calcium carbonate in the laboratory. Various blue-green algae, or cyanobacteria, such as *Synechococcus elongatus* and *Gleocapsa* sp. are also constituents of moonmilk. In the darkness of the cave, they must be able to feed by means of metabolic pathways other than the normal photosynthetic ones they use in the light.

A quite different, but equally striking bacterial phenomenon in caves is the ebony black patina of manganese minerals which forms on ancient bones in Mendip caves and sometimes on calcite flows and round the top of the margins of rimstone pools. There are beautiful examples of this coating in Otter Hole and Ogof Ffynnon Ddu.

The great pioneer in unravelling the composition and role of micro-organisms in caves was Endre Dudich, who published his most significant work during the 1930s. Since then, important advances have been made by Victor Caumartin (1950s) and A.M. Gounot (1960s) working in France's national underground laboratory at Moulis-Ariege. Gounot specialized in the bacteria of cave silts, studying their population dynamics, nutritional requirements and physiological activities in some detail. She found population densities of the order of several million to several hundred million bacteria per gram of cave silt, not far off the density expected in a good agricultural soil. Using the technique of thin-film chromatography, she found that bacteria were enriching the silt with a range of complex organic substances such as amino acids and vitamins, and experimentally showed that *Niphargus* species depend for their successful development and growth on a factor (possibly a B-group vitamin) produced by bacterium.

Different bacteria may feed either autotrophically (synthesizing food from inorganic chemicals), or more usually, heterotrophically (depending on ready-made organic material). The latter group includes important pathogens and agents of decay. They are present in soils, water, the air and indeed all over the surface of the earth and therefore also in caves. As this is not a textbook of microbiology, I propose, with but a single exception, to say no more about those species found in caves except that the list is a long one.

The exception I shall make is in respect of one particularly unpleasant pathogen which is much feared by cavers. I refer to the spirochaete, *Leptospira interrogans*, the causative agent of Weil's Disease, a condition which may cause serious damage or complete failure in the liver and kidneys of sufferers, with fatal results if treatment is sufficiently delayed. *L. interrogans* lives in a wide range of wild animals, particularly rodents, but even amphibians and reptiles have been proved to be carriers. There are about 180 different strains (or serovars) of the bacterium, which are defined by proteins (antigens) on their surface, to which antibodies may be raised in those who have had exposure to them. The predominant infecting serogroups found in the British Isles are *icterohaemorrhagiae* (carried by the brown rat), canicola (carried by dogs) and *sejroe* (carried by cattle). The most serious infections are due to *icterohaemorrhagiae*. In a recent survey, 9% of a sample of British cavers were found to possess antibodies demonstrating some contact with the pathogen, and I know of two cases of Weil's Disease contracted by Mendip cavers as a result of visits to Stoke Lane Slocker (a cave much frequented by brown rats) and GB Cave. The use of caves as convenient rubbish dumps by farmers and others no doubt contributes to their attractiveness to rats, and so to the danger faced by the cavers who visit them.

Chemo-autotrophic bacteria are relatively more common in rocks and subterranean voids than on the earth's surface and so have received considerable attention from cave microbiologists. They include the Iron Bacteria, such as *Ferrobacillus ferrooxidans*, which is widespread in mines, particularly in bituminous coal deposits, and in caves. It is a strict autotroph using carbon dioxide as a carbon source and obtaining its energy by oxidizing ferrous iron to ferric sulphate or ferric hydroxide. An interesting iron autotroph called *Perabacterium spelei*, of curious morphology and uncertain taxonomic position, was isolated from the sediment of caves in southern France by Caumartin. It is rod-shaped, feeds solely on inorganic compounds, is capable of anaerobic growth and may prove to be confined to subterranean habitats.

The Sulphur Bacteria fall into two groups: those which oxidize and those which reduce sulphur compounds. The former category includes *Thiobacillus thiooxidans*, which derives energy by oxidizing elemental sulphur or thiosulphate, and *Thiobacillus ferrooxidans* which does the same, but will also oxidize ferrous to ferric iron autotrophically. This ferric iron is then capable in turn of oxidizing metal sulphides, such as iron and copper pyrites to the sulphates, and *T. ferrooxidans* is used industrially on a large scale in concentrating metals in low-grade or inaccessible ores, or mine wastes. Most bacteria can reduce sulphur compounds to a limited extent, but *Desulphovibrio* spp., which are widespread in caves, derive their main energy requirements by a simultaneous anaerobic reduction of sulphate and oxidation of organic matter to produce carbon dioxide and hydrogen sulphide.

Another group of autotrophic bacteria metabolise nitrogen compounds. Nitrifying bacteria get energy by oxidizing ammonia to nitrite (nitrosification) or to nitrate (nitrification), using carbon dioxide as their sole carbon source. Nitrosifying bacteria include *Nitrosomonas* and *Nitrosococcus*, both found in South Wales caves, often in the water of small pools. Nitrifying bacteria include *Nitrobacter*, also recorded underground in South Wales, like some lichens and fungi, they are able to dissolve limestone. They do this by oxidizing ammonia to nitrous and nitric acids which react with the limestone to form carbon dioxide and soluble nitrite and nitrate ions. Apparently mines in the Russian brown-coal deposits near Moscow all contain methane-forming bacteria, which generate the gas solely from carbon dioxide and hydrogen, under aerobic conditions.

The final group of possible autotrophs are the Gram-negative aerobes (recognizeable in microscope preparations by their reaction to staining) which, when supplied with a 'start-up' source of chemical energy, can fix atmospheric nitrogen, converting it to organic nitrogen compounds. Of these, *Azotobacter* sp. has been recorded from a Yorkshire cave, while both *Clostridium* and *Azotobacter* spp. have been found in caves in South Wales.

Fungi

Fungi possess no chlorophyll, and feed heterotrophically in much the same way as animals, collecting ready-assembled organic molecules from the tissues or detritus of other organisms. Fungi grow so well in damp, dark places, that as early as the 17th century, commercial mushroom farmers were siting their operations in caves near Paris and by the late 19th century it is estimated that 2400 km of cave space was devoted to mushroom cultivation. This being so, one might expect caves to be filled with a colourful array of fungi, jostling each other for space on the mud banks of subterranean streams. In fact, the characteristic 'toadstool' fruiting bodies of the basidiomycetes, or 'higher fungi', are quite a rare sight in caves, though fungal hyphae – the threads with which fungi feed – are commonly seen. The reason for this state of affairs is, predictably, to do both with the nature of the food supply, and with the absence of light. Higher fungi tend to grow only on fairly concentrated sources of organic material – dead or living tree trunks, accumulations of leaf litter, or animal droppings. In caves, organic detritus tends to occur as thin deposits rather than as substantial dollops, and the larger fungi may simply not be able to concentrate enough energy and therefore build enough tissue to form a fruiting body. The problem is compounded by the perpetual darkness and relative constancy of the cave environment.

An interesting physiological adaptation of fungi, which has not as far as I am aware been confirmed in caves, is their ability to render soluble certain rocks when growing under conditions deficient in essential inorganic ions. This characteristic is of particular importance to the lichens, which are associations between fungi and algae. Lichens have been recorded in the dark zone of tropical caves, but not so far from caves in Europe.

Fungi may enter the cave, or mine, in one of three ways: their spores may be carried in on air currents, by visiting animals, such as insects, bats or cavers, or already attached to mine timbers or water-borne detritus. Most mine timbers are softwoods, so the fungi found in mines tend to be species associated with softwoods.

Fig. 4.19 *Heleomyza* fly sprouting the fruiting bodies of the insect-eating fungus *Hirsutella dipterigina*.

The life cycle of a fungus begins when a microscopic spore arrives on a suitable food material. The spore germinates and grows a threadlike hypha, which branches repeatedly, secreting enzymes onto the food substrate and absorbing back the products to fuel further growth. Given adequate nutrition, the mass of hyphae forms a mycelium which may have a characteristic appearance. Fungal enzymes are tailored to digest the complex molecules of a particular food. In wood-rotting species, specific enzymes attack either the cellulose or lignin (the substances which respectively give timber its tensile and load-bearing strengths) so that the wood rots in a particular way, depending on the species of fungus growing on it. As the food source is consumed, the fungus puts its accumulated resources into a mass of specially modified hyphae which grow away from the substrate to form the familiar cap-, or bracket-shaped fruiting body. This will have a a series of gills, generally on the underside, which produce and shed millions of minute spores. The spores are very light and can be wafted by the slightest air current to the next source of food to start the cycle once again.

Higher fungi are more often encountered in mines, especially in abandoned workings, where pit props and other timbers supply an appropriate source of food in sufficient quantity. On the surface, fruiting bodies are produced in an annual cycle, timed in response to changing combinations of daylength, temperature and humidity. In the absence of these cues, the subterranean fungus follows a cycle mediated by its own metabolic clock, but over time this becomes increasingly erratic. Some species fruit sporadically, but energetically, over several months until seemingly exhausted and then slump into a long period of quiescence. Others seem to fruit only when stimulated, particularly after they have been disturbed; a timber falling or being moved is often followed by a flurry of fruiting bodies.

Some fungi do not rely solely on producing spores to spread themselves, but instead may send out thick adventitious hyphae called rhizomorphs as their food source is used up. In mines these have been traced from one infected prop to another across 50 m of floor. The commonest species that does this is the Bootlace, or Honey Fungus *Armillaria mellea*, and it is also seen in Candle Snuff Fungus *Xylaria hypoxylon* and Dead Man's Fingers *Xylaria polymorpha*.

The fruiting bodies produced by subterranean fungi are often strangely atypical and may be sterile. Close up, the colours can be unexpectedly brilliant and intense, with red, white and yellow predominating. The same species may vary depending on the particular type of wood on which it is growing, or even according to the type of preservative used to try to retard fungal growth. Worse still, some species may fail to produce any kind of fruiting body in total darkness and may instead develop a monstrous vegetative mycelium covering several square metres. As the shape, colour and size of the 'cap', and the microscopic appearance of the spores are the characters generally used to identify fungi to species, the erratic and bizarre forms encountered underground are often quite unrecognizeable.

Despite the problems with identification, it is possible to list some of the commoner species encountered on mine timbers. The principal cause of decay for pit props in coal mines is *Poria vaillanti*, which is also found in caves, where it may spread a fan-shaped sheet of hyphae across large areas of the cave floor, or form huge, hanging draperies of tasselled mycelium. Other wood-rotting fungi found in mines include *Lentinus lepideus*, which often forms abnormal cylindrical fruiting bodies; the Dryad's Saddle, or Scaly Polypore *Polyporus squamosa*; the Dry Rot fungus *Merulius lachrymans*; Hoof or Tinder Fungus *Fomes fomentarius* and *F. annosus*; several species of polypores – *Coniophora puteana*, the major cause of Wet Rot, *Paxillus panuoides* and *Lenzites betulina*, in addition to the species referred to earlier. Most of these species are well documented because of their economic significance in attacking pit props and other structural timbers in mines, and considerable research has been conducted into ways of controlling them by means of fungicide chemicals.

The Inkcaps (such as *Coprinus domesticus*) are common in caves as well as mines, where they produce dwarfed, but otherwise normal and fertile fruiting bodies. It seems that *Coprinus* species are particularly well adapted to underground life and they are widely reported from caves as far afield as Malaysia and the USA. The commercially grown Field Mushroom, *Agaricus campestris*, sometimes shows up in mines where horses have been used. Agarics in caves often have very elongated stipes (stems) and Dobat has described a cave-frequenting *Mycena* species which had hanging stalks up to 20 cm long, turned up at the ends so that the caps were held horizontally to enable spores to be shed from the gill lamellae.

In addition to the obvious large fungi found on mine timbers, caves may contain a whole host of other microfungi which can be studied only under a microscope. They may be isolated from the soil, air and water of the cave by agar plating and other microbiological methods. Fungi Imperfecti of the genera *Penicillium* (such as *P. italicum*) or *Aspergillus* (such as *A. versicolor*) are commonly found growing saprophytically on all kinds of organic debris underground. They can survive long periods as mycelial cysts, formed in response to the presence of sulphide ions in cave sediments. Sulphides are often liberated by the bacteria present in wet cave clays, which are thus able to suppress the fungi which might otherwise compete for limited nutrients. In the presence of ferric salts, the cysts quickly sprout mycelia, and these may produce chemicals which in turn inhibit the growth of bacteria. Cubbon has

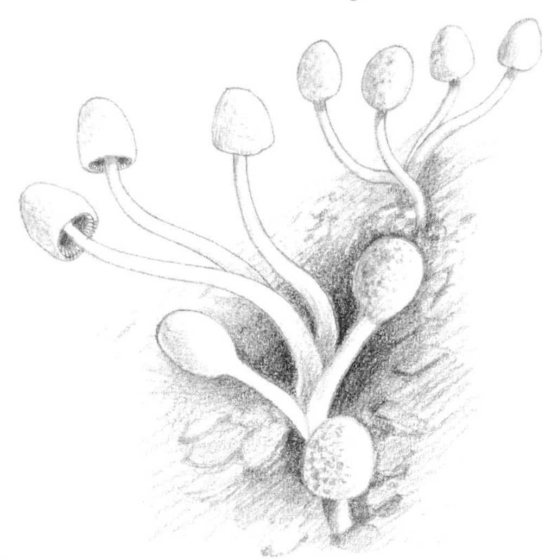

Fig. 4.20 The Ink-cap fungus *Coprinus domesticus* produces characteristic fruiting bodies underground which look quite different to its normal above-ground growth form.

shown that the Phycomycete *Absidia repens*, collected from a cave, was able in the laboratory to produce an antibiotic active against the bacterium *Staphylococcus aureus*, and Hennebert, working in a Belgian cave, demonstrated the production of an antifungal antibiotic by *Oidiodendron* sp. which was active against *Cladosporium herbarum*, isolated in the same cave.

Some Fungi Imperfecti, a group characterized by the absence of a sexual stage of reproduction, are nowadays considered to be the imperfect stages of other 'higher' Ascomycete or Basidiomycete fungi. One such is *Oedocephalum lineatum*, a species originally described from the galleries of wood-boring beetles, but which also occurs as a wall fungus in the Bristol Waterworks Heading at Burrington Combe on Mendip. *O. lineatum* is now recognized as being the imperfect conidial (a type of asexual spore-propagating) stage of *Fomes annosus*, a Basidiomycete which is common on decaying wood above and below ground.

Fungi in the group Agonomycetes have never been observed to form asexual spores. "*Ozonium auricomum*", a fungus originally ascribed to this group, has been reported from many caves and mines, where it appears as a mass of rather coarse, reddish-orange threads. However, more recent observations suggest that it is in fact simply a sterile vegetative stage of *Coprinus* sp. (a Basidiomycete), groups of which have been observed arising from orange cushions of the "*Ozonium*" in Wookey Hole, Somerset. I have also seen the same fungus two kilometers into the fossil trunk passage of Otter Hole, near Chepstow.

One organism from Wookey Hole has defied all attempts to classify it. It was isolated from a reel of cable submerged in the stream passage, where it was growing saprophytically on the cable's insulation material. A detailed investigation of its structure and nutrition was undertaken by Willis (1961), during which the organism showed a great plasticity of form. It appears to have affinities with both the algae and the fungi, and did not form any recognizable sporing structures.

Cavers often report observations of 'fungal hyphae' in underground streams. Most of these are probably filamentous chlamydo-bacteria, but some may be common aquatic Phycomycetes such as *Achlya* or *Saprolegnia* which can produce very thick hyphae. The presence of 'water fungus' may be correlated with the occurrence of aquatic cavernicoles, such as the cave water louse *Asellus cavaticus*.

The Myxomycetes, better known as Slime Moulds, are bizarre, often brightly coloured fungi which, during the feeding stage, form an amoeba-like plasmodium which creeps about in search of food. When mature, the plasmodium stops its wandering and differentiates into a series of spore-filled capsules. These generally have some mechanism for hurling their spores out into the air in order to increase their chances of spreading to a new source of food. The Slime Mould *Trichia favoginea* is relatively common in the Box Freestone Mines in Wiltshire.

Fungi are important in the cave ecosystem in several ways. They play a fundamental part in recycling nutrients and making available organic nitrogen compounds, vitamins, and so on, to the cave community, many of whose members are adapted specifically to feed on fungal hyphae and fruiting bodies. The traffic is not all one way, however, for certain fungi themselves act as predators or parasites on cave invertebrates.

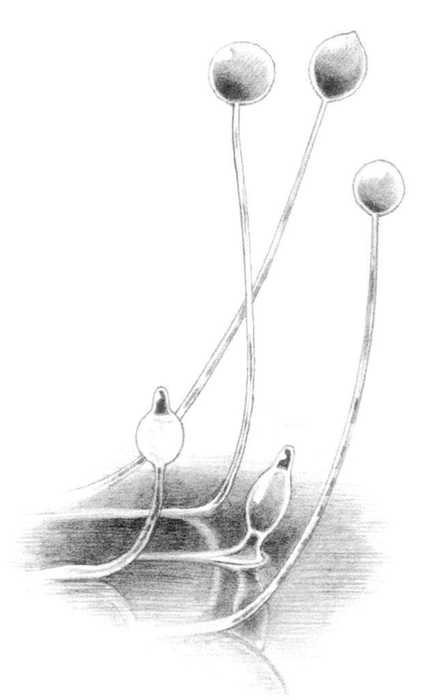

Fig. 4.21 Fruiting capsules of the slime mould *Dictyostelia mucoroides*.

According to Cubbon (1976), there have been no identifications to date of nematode-trapping fungi, such as *Dactylella* or *Arthrobotrys* spp. from caves. This is surprising, as nematodes (roundworms) abound in wet cave sediments and many of the fungi that prey on them live in similar conditions above ground.

A fair number of fungi in caves are associated with the bodies of cavernicolous invertebrates. Some, such as *Paecilomyces* (recorded in caves in Yorkshire and South Wales), are simply saprophytes which grow on the remains of dead insects, but others, such as *Beauveria* and *Hirsutella* spp. attack live invertebrates, causing disease and death while completing their life cycles in the tissues of their host. *Beauveria* is associated particularly with dipterous flies in caves. Its spores have been shown to require at least a 90% relative humidity for successful germination, and to grow as well in darkness as in the light. The entomophagous species *Hirsutella dipterigina* in Britain is closely associated with caves and in particular with the fly *Heleomyza serrata*. The fungus-ravaged carcasses of *Heleomyza* present a bizarre appearance, with a forest of the twisted, antler-like fruiting bodies of the fungus poking out in all directions. Another fungus, *Stilbum kervillea*, has been thought to be hyperparasitic on *Hirsutella*, but if so it is a facultative association, as the two species are often found separately. Entomophagous fungi, such as *Hirsutella*, and parasitic Laboulbeniales may be important in controlling the densities of individual species of arthropods in the cave.

A number of fungi are known exclusively from caves. They include saprophytic species, such as the Ascomycete *Lachnea spelaea* from Italy, but also highly specialized members of the order Laboulbeniales which are obligate parasites on individual species of troglobitic arthropods. The great majority of species live on beetles, but a few are known to attack mites. Infections are not necessarily lethal, and the living insects can be seen running about with a number of the tiny, club-shaped fruiting bodies of these fungi protruding from their integument. *Laboulbenia subterranea* and species of *Rhachomyces* are frequent parasites of cavernicolous *Trechus* beetles in Europe, but I do not know whether they have been recorded on our own *Trechus micros*.

The association of fungi with cave faunas is not confined to invertebrates alone. The dermatophytic fungi which cause skin lesions of the 'ringworm' type in man are known to affect bats. Kajihiro has isolated ten species, all potential human pathogens, from bat guano in various caves in the USA. *Trichophyton persicolor* has been recorded on a British Pipistrelle bat and a fungus tentatively identified to the same genus has been found in a cave in Yorkshire.

Cavers frequently report seeing colonies of 'wall fungus' in British and Irish caves. These are off-white, lichen-like patches, typically 1–3 cm across, sometimes with a yellow centre and often beaded with droplets of condensation or exudate which pick up the beam of a head-torch, making them very conspicuous. They are most frequent in the 'transition zone' at and just beyond the limits of light penetration, and are largely responsible for the characteristic earthy smell of caves and mines, known as 'cavers' perfume'. Mason-Williams and Holland have shown that the dominant organisms present are Actinomycetes of the genus *Streptomyces*, mixed in with with a fungus, *Fusarium*. It is thought that there may be an intimate biological association between the two, as mixed cultures in the laboratory produce a similar growth form to cave colonies. Actinomycetes are intermediate between fungi and bacteria and are known mainly for their ability to produce antibiotics. We shall examine the possible role and implications of such substances in the life of the cave in the next section.

Protista

Free-living protists are common in caves and seepage waters, but have barely been studied. If a sample of water and silt from the bottom of a cave pool is examined under the microscope, ciliates can often be found, and more careful searching sometimes reveals the beautiful quartz-grain case of *Difflugia* (Sarcodina). The protists are not the easiest animals to study, but anyone with patience and a microscope could make a useful contribution in this neglected field.

Porifera and Bryozoa

Sponges seem never to have been recorded from British caves and only very rarely from any others. This is surprising, since fresh-water sponges such as *Spongilla lacustris* have often been reported encrusting the insides of water pipes, as also have certain Bryozoa or moss animals – another group almost unrecorded from caves. There is no obvious reason why either of these should not occur in gently-flowing underground streams and it would be worth while to look out for them.

Turbellaria

Some species belonging to the order Rhabdocoela are interstitial and a few, unrepresented in Britain, are cavernicolous. The order Tricladida includes marine and terrestrial forms, but it is the fresh-water triclads or 'flatworms' which are most familiar, and this is the group which is particularly rich in troglobites in continental Europe, North America and Asia. Flatworms are not common in British caves and most of the species recorded are probably accidentals except for *Phagocata vitta, Crenobia alpina, Polycelis felina* and *Dendrocoelum lacteum* which are sometimes cavernicolous. All four species are usually found in surface habitats. The first two are associated with cool waters, either at high altitude, or in spring-heads, as well as in most cave areas in Britain. *Phagocata vitta* is the only flatworm known to inhabit Irish caves, where it has been recorded from Clare, Sligo and Fermanagh. In Britain, *Polycelis felina* seems to occupy similar habitats to *Crenobia alpina*, and the two species may influence one another's ditributions through competition (Beauchamp & Ullyott, 1932). *P. felina* has been found in a well in Banham, Norfolk which also contains a large population of the troglobitic amphipod, *Niphargus kochianus*.

According to diver Rob Palmer, *Dendrocoelum lacteum* is common in the phreatic sections of the Cheddar River beneath Gough's Cave and the River Axe in Wookey Hole, which also support a large population of unusually large *Asellus cavaticus* isopods. *D. lacteum* is fast-moving, with a shallow anterior sucker, both features which enable it to capture live prey, especially *Asellus* (which shares an identical distribution) with great facility. With its ability to breed in temperatures lower than any other British species, and its fondness for *Asellus, Dendrocoelum* could be described as being preadapted to cave life, and indeed it has many troglobitic relatives in Europe, which are eyeless and almost transparent.

Fig. 4.22 The flatworm *Crenopia alpina* is particularly associated with mud-floored, sluggish underground streams.

Nematoda

Free-living nematodes can often be seen wriggling in their characteristic way in samples of cave water and silt examined under a low-power microscope. These animals are notoriously difficult to identify, so species records are few.

Annelida

Among the Annelida, or segmented worms, the polychaetes are almost entirely marine, but at least four interesting species are known from underground fresh waters in central and eastern Europe, the Balkans, Japan and the highlands of Papua-New Guinea. They are generally thought to be ancient marine relicts, possibly having left the sea in Tertiary times.

The oligochaetes are mostly terrestrial and fresh-water forms, and this is the class to which the majority of cavernicolous annelids belong. Troglobitic oligochaetes, both aquatic and terrestrial, are known from various parts of the world, but all those identified so far from British caves are also known from non-cave habitats. The commonest and most widespread species recorded are the lumbricids *Eiseniella tetraedra, Allolobophora chlorotica, Dendrobaena rubida, D. subrubicunda* and *Bimastus tenuis*. All of these are able to maintain permanent populations in cave passages where the mud in which they live is replenished from time to time by organic material deposited by flood-waters. Even so, the utilizable content of such sediments may be so low that cavernicolous earthworms may be close to starvation for much of the time (Piearce, 1975). At least one enchytraeid (pot-worm), *Enchytraeus bucholzi* recorded occasionally in British caves is a widespread and frequent inhabitant of European caves, turning up in and out of water and even in rimstone pools in the deepest parts of caves. It is probably more or less specialized to life in mesocaverns and should certainly be regarded as a cavernicole.

Aquatic oligochaetes are mostly very small worms which are easily overlooked, and they have tended to be neglected. An isolated population of *Aelosoma hemprichi* living in a drip-fed pool in Ogof Ffynnon Ddu in South Wales, was observed by Jeff Jefferson and myself at intervals from 1972 and 1978. The species is transparent and colourless except for the contents of scattered epidermal cells which embellish it with bright orange-red spots. *Aelosoma* is able to reproduce asexually by a process of fission, and it seems that the population was able to maintain itself for several years (and presumably a number of generation-s), feeding on the silt of the pool bottom. A number of cavernicolous leeches are known from other parts of the world, but there is no evidence that those found in British caves are able to maintain permanent populations underground.

Mollusca

Within the Mollusca, only two classes, the Bivalvia and the Gastropoda, are found in freshwater caves. Very few bivalves of any kind have been found in British caves, although Moseley (1970) reported finding numerous specimens of the tiny *Pisidium personatum* in Hazel Grove Main Cave in the Morecambe Bay area. As these formed an isolated group, surrounded by an apparently adequate food supply of rich peaty mud, it is possible that they may have formed part of a viable cavernicolous population. Both *P. personatum* and *P. obtusale* are commonly found in Irish caves, often under muddy conditions too foul for other species.

The Gastropoda are much better represented in caves. Numerous troglobitic species, both terrestrial and aquatic, are known from many parts of the world. The aquatic cavernicoles are often tiny species within the family Hydrobiidae, frequently adapted to life in the mesocaverns and common almost everywhere on the continent of Europe. Lacking our own native cavernicolous hydrobiids, British and Irish caves would seem to present a ready-made unexploited habitat for any hydrobiid species which can make the required adaptive shift, and it seems that in Ireland, one such species has indeed begun to move into caves. This is the ecologically adventurous *Hydrobius jenkinsi*, recorded from the Doolin Cave system and the Poulnagollum-Poulelva system, both in Co. Clare. This snail has a strange history. It appeared suddenly in the mid 19th century as a brackish water species and started moving into fresh waters in vast numbers in 1893, the same story being repeated all over Europe. As it gives birth to live young, rather than laying eggs, and can reproduce without mating, the species would seem equipped to thrive in caves, provided it can find a suitable food supply. The colonies in the previously-mentioned caves seem to be well established, although their shells have developed peculiarly distorted whorls which make them quite difficult to recognize.

Most terrestrial cavernicolous snails belong to the family Zonitidae, notable for their omnivorous tendencies. Many species produce chitinase enzymes which are able to digest the tough proteinaceous exoskeletons of insects. *Oxychilus cellarius* seems to be quite common in caves in Britain and Ireland and may be a cavernicole, feeding on the bodies of insect members of the 'wall association' (see next chapter), although the evidence for this is rather anecdotal. A number of other related snails have been recorded in British caves such as *Oxychilus draparnaldi, O. alliarius, O. helveticus, Retinella nitidula* and *Zonitoides nitidus*, but these are not known to be cavernicolous.

Arachnida

Our largest and best known cave spider is the tetragnathid *Meta menardi* which can reach a body length of 16 mm and a leg-span of 50 mm. It spins a relatively small orb-web which requires little space, so that the spider seems equally at home in mesocavernous spaces as in caves, where, together with its threshold-dwelling cousin *M. merianae*, it forms part of the 'wall association' discussed in the next chapter. *M. menardi* occurs throughout the British mainland, Europe, Asia, Japan, North America and even Madagascar. Curiously, *Meta* spiders are seldom recorded in Irish caves, and indeed spiders of all sorts are far less often reported from Irish than from British caves.

The linyphiids, or money spiders, include a large number of small species which live in leaf litter, soil, animal burrows and caves. One genus in particular is associated with subterranean life: *Porrhomma* spiders are about two millimetres long and have a northerly distribution right up into the arctic circle. They include a number of species which are considered to be cryophilic relicts, left behind in suitable cool and sheltered habitats at the end of the last glacial era. Four of these are associated with underground habitats in Britain and Ireland. The blind, straw-coloured *Porrhomma rosenhaueri* is a troglobite known from several European caves. In Britain, it has been found only in South Wales, and then only in two caves: Lesser Garth and Ogof-y-Ci. It is perhaps better established in Irish caves, having been recorded from Mitchelstown New Cave in Co. Tipperary and Fisherstreet Pot in Co. Clare. *P. egeria* is a pale species with greatly reduced eyes which is frequent in caves, cellars,

Fig. 4.23 The blind cave spider *Porrhomma rosenhaueri*.

mines and under stones. Its principal habitat is probably in the SUC (the complex of mesocavernous rock spaces immediately below the soil). *P. campbelli* is a very rare species which has been found occasionally under stones and in a mole's underground nest. *P. convexum* is a very common species found in caves, cellars, mines, under stones and also in houses and vegetation undergrowth. Specimens taken outside caves are very dark brown with a greyish black abdomen. Those in underground situations may be brown or quite a bright orange, with a white abdomen. *P.convexum* is common in Mendip and Yorkshire caves and there are records from Derbyshire and the Forest of Dean, but it seems to be quite rare in the caves of South Wales and Western Ireland where *rosenhaueri* occurs. Apart from the latter's skinnier legs and regressed eyes, these two species are very similar in the configuration of spines and trichobothria (sensory hairs) on their legs and in their genital structures, suggesting a close kinship. It may be that *P. convexum* is the ancestor of *P. rosenhaueri* and indeed the spiders were originally classified within the same species, being known respectively as *P. proserpina* and *P. proserpina myops*. Two other members of the genus *Porrhomma: pygmaeum* and *pallidum*, though not often found in caves, are considered by some spider specialists to be able to maintain subterranean populations, as are two other linyphiid spiders, *Lessertia dentichelis* and *Lepthyphantes pallidus*.

In France, several spiders (including *Porrhomma egeria*) have been collected in the SUC rock-crevice habitat. Although this habitat has received little study in Britain, at least three spiders have been recorded in situations which suggest that they may be SUC, or mesocavern specialists. *Centromerus persimilis* is a pale, cryptic little money spider known in Britain from just two specimens, one of which (a male) was collected from the crevice of a limestone pavement at Malham Cove in Yorkshire in 1969. More recently, Rowley Snazell of the Institute of Terrestrial Ecology has captured two most interesting money spiders in pitfall traps set into the turf of a chalk hillside at Lyscombe Hill in Dorset. *Hahnia microphthalma* is a pale yellow spider with reduced posterior median eyes. *Pseudomaro aenigmaticus* is again pale yellow in colour, and has more markedly reduced eyes (the small anterior medians remain pigmented; the others are unpigmented and indistinct). Snazell suggests that both species may inhabit the network of small fissures and solution channels found in the chalk subsoil on the site.

Surprisingly few harvestmen have been recorded in subterranean habitats in the British Isles. The small, cryptic *Mitostoma chrysomelis* seems to be about the only species at all likely to maintain permanent populations in caves, although this has yet to be confirmed.

The list of terrestrial mites recorded from British caves is extensive and includes a number of probable cavernicoles, and even a few troglobites. Among the Mesostigmata (mites), *Eugamasus magnus* and *E. loricatus* are widespread in underground situations. The latter has only rarely been found anywhere other than in caves. *E. magnus* is at the centre of a species complex – a whole range of closely related forms, some no doubt specifically distinct, others merely subspecies or races – which requires taxonomic revision. Two members of the complex are considered to be troglobites. One of them, *Eugamasus traghardi*, was originally recorded from the Grotte d'Istaurdy in France as a variety of *E. magnus*, but was later accorded specific rank. The other is *E. anglocavernarum*, first found by E.A. Glennie in Bagshawe Cavern, Derbyshire in 1938 and described as a new species by F.A. Turk. This mite has subsequently been found again in Bagshawe Cavern and in a few other underground localities in various parts of Britain, but the number of records is small. Other mesostigmatid mites frequently recorded from British caves include species of *Veigaia*, one of which, *V. transisalae*, appears to be cavernicolous.

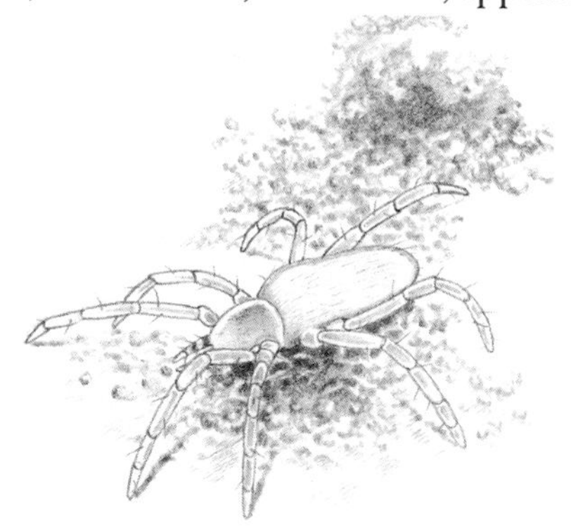

Fig. 4.24 The fast-moving *Rhagidia spelaea* is one of the commonest mites in British and Irish caves.

The other group of mites in which cavernicolous forms are particularly well represented is the Prostigmata, and among these the genus *Rhagidia* stands out in Britain. *R. spelaea* is one of the commonest mites in British and Irish caves and is a cavernicole, as also is the somewhat rare *R. gigas*. F.A. Turk, who has studied these creatures in detail suggests that *R. spelaea* may actually represent a species complex with a number of distinct forms represented particularly in Irish caves, but in the absence of agreement over the taxonomic significance of the species' most variable characters, for now the matter has been left in abeyance.

Another uncommon species, *R. longipes*, is a troglobite and this status may also be accorded to two other rare species, both discovered in the late 1960s during the heyday of faunal collecting in British caves. *R. odontochela* was described by Turk from a single female specimen found by W.G.R. Maxwell in Ogof Cynnes; and *R. vitzthumi*, from two females collected by Dr Pat Cornelius in Rift Cave, Devon, and again by W.G.R. Maxwell in Spratts Barn Mine, Oxfordshire. During 1968 the same two workers also collected one specimen each of a third new and possibly troglobitic prostigmatid mite, since named *Bonzia brownei*, from Rift Cave and Reed's Cave, Devon. I mention these individual discoveries to illustrate just how valuable a contribution has been made to our knowledge of cave faunas by motivated amateur collectors such as W.G.R. Maxwell, who must have chalked up quite a few new species during his active years. I am sure that the same opportunity still exists today for anyone with the eyesight and motivation required to mount a systematic study of underground populations of these tiny and taxonomically difficult beasts.

Some of the terrestrial mites found in caves are closely associated with water and a few may well prove to be substantially aquatic in habit. However, the true water mites (Hydracarina) are seldom found in caves, and this is rather surprising since many forms are known to inhabit superficial phreatic waters. Two blind and depigmented members of the family Porohalacaridae, *Soldanellonyx monardi* and *S. chappuisi* have been recorded from caves in Yorkshire. They are probably cavernicoles, although they also occur in superficial groundwaters and lakes. Jefferson (1976) comments that the sparsity of cavernicolous water mites in Britain is probably more apparent than real and a search in the right situations would probably yield more, but it is curious that relatively few have been reported from caves in any part of the world.

The parasitic acarines (mites and ticks) occurring in caves in Britain and Ireland are those associated with bats. Much the most common is the tick *Ixodes vespertilionis* which can infest any of our native bats. This species and others were discussed earlier in the section titled *A place of shelter*.

Crustacea

The Crustacea are largely marine, but there are many freshwater and some terrestrial groups which contribute to the cave fauna. Worldwide, highly evolved troglobitic Crustacea abound in almost every region of the globe and form perhaps the most important Class underground. Many troglobitic species belong to archaic groups which show relictual distributions, and these include representatives of no less than four orders which are largely confined to subterranean waters. Although the Crustacea are well represented in the caves and groundwaters of these islands, the contributions of its various groups

Fig. 4.25 The ostracod *Cypridoptis subterranea* in Britain appears to be confined to groundwaters and caves.

are curiously uneven, and only three subclasses, the Ostracoda, Copepoda and Malacostraca, are of interest to the cave biologist in Britain and Ireland.

The Ostracoda are small crustaceans, either marine or fresh-water, with a bivalved carapace: easy to recognize as ostracods but difficult to identify. Some of them swim while others crawl, and, particularly among the latter, there are many interstitial forms including some which occur in the deeper cave waters. Many ostracods are found in springs, and for some this may be their normal habitat, while others, such as *Eucypris anglica* which appears in bourne risings (where water boils up from deep underground), are almost certainly cavernicoles. Perhaps the first record of ostracods in British caves was that of Lowndes, who collected two species from the Corsham stone mines in 1932, *Herpetocypris palpiger* and *Candona wedgewoodii* – both now considered to be troglobites. There are many more troglobitic species of *Candona* on the continent and several of these occur in subterranean waters in Britain and Ireland, although the systematics of the group will need sorting out before an exact list can be produced. *Cypridopsis subterranea* is quite numerous in the mesocavernous seepage water running over flowstone slopes in caves such as Ogof Ffynnon Ddu in South Wales and is otherwise known only from springs.

A fair range of copepods has been recorded from British and Irish caves, and they are often present in wells and other groundwater sources. None of those identified so far appears to be troglobitic, but several, such as *Paracyclops fimbriatus* and *Acanthocyclops vernalis* (known from caves in South Wales and Mendip), *A. viridis* (as above, but also Derbyshire, Yorkshire and Fermanagh) and *A. venustus* (found in the Poulnagollum-Poulelva system in Clare), are widespread cavernicoles. Since there are troglobitic copepods in other parts of the world, and our own fauna of smaller subterranean crustacea has been little studied so far, some may still be found here.

Among the Malacostraca, the subclass containing most of the larger and more familiar crustaceans, only the Syncarida, Isopoda and Amphipoda are represented in the British underground fauna, but elsewhere there are also cavernicolous mysids and decapods.

The syncarid *Bathynella stammeri* has been found in seeps in a Wiltshire stone Mine and in two caves in Yorkshire. It is also found quite abundantly in interstitial ground waters in Devon and the Thames Valley and its small body size (1 mm long and 0.1 mm in diameter) would suggest that it is primarily an interstitial troglobite (and may also account for its presence being frequently overlooked).

Only one troglobitic isopod is known from Britain. This is the small aquatic *Proasellus cavaticus*. It was first recorded in Britain from a well in Hampshire, and is occasionally found in springs and gravels, but in its main areas of distribution (South Wales and Mendip) it is essentially cavernicolous, occurring particularly on the underside of stones in underground streams and in the film of water flowing over stalagmite slopes. There appear to be two distinct forms, of very different sizes. The smaller (up to four millimetres long) lives in the unsaturated zone of Mendip caves, while the larger (up to eight millimetres long) is found in caves around the Welsh coalfield, in Otter Hole near Chepstow, and in the phreatic passages of Cheddar River Cave and Wookey Hole. Dr E.M. Sheppard, who has examined many specimens, is inclined to the view that the two forms may represent different subspecies, and Jefferson (1989) reports that the larger form may produce more young per brood.

The common river-dwelling isopod, *Asellus meridianus*, is often found in swallet caves, and in some cases it is possible that they maintain permanent underground populations. Guzzle Hole on South Wales' Gower Peninsula apparently contains "aberrant" *P. meridianus* (Jefferson, 1989).

Among the terrestrial isopods recorded from British and Irish caves, one, *Androniscus dentiger*, is certainly a cavernicole, while a further four, *Oniscus asellus*, *Cylisticus convexus*, *Haplophthalmus danicus* and *Trichoniscoides saeroeensis*

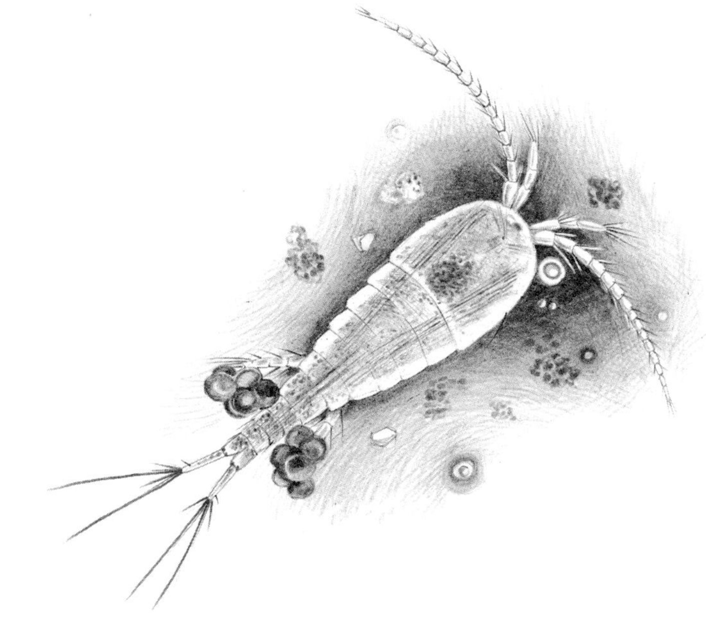

Fig. 4.26 The copepod *Acanthacyclops viridis* is widespread in groundwaters in Britain and Ireland.

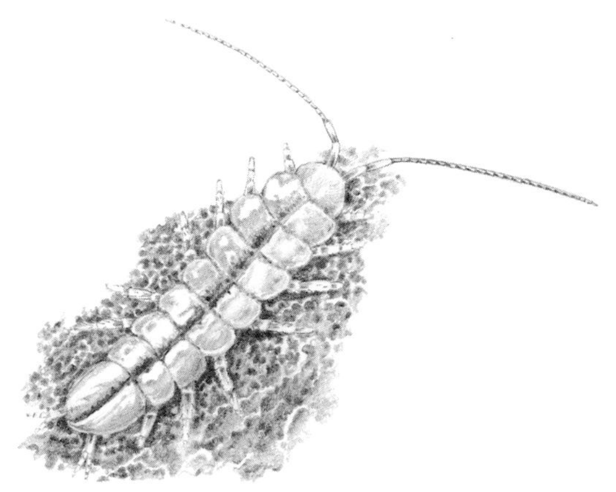

Fig. 4.27 The eyeless, depigmented cave isopod *Proasellus cavaticus*.

may maintain populations occasionally in caves. *A. dentiger* is a pretty little pink woodlouse which seems almost as much at home in water as on land, and is frequently recorded under stones, as well as from caves and mines (although it is rare in Irish caves) usually fairly close to the surface, which makes me suppose that it is essentially an inhabitant of 'SUC' mesocaverns. *C. convexus* has somewhat similar habitat preferences, and is recorded from tunnels and underground drains. The tiny, blind *T. saeroeensis* (less than three millimetres long) was first recorded in Britain and Ireland from caves (Morecambe Bay in 1964, County Clare in 1968) where it must surely be able to maintain permanent populations, but subsequent records have shown it to be essentially a coastal species.

In Britain, the amphipods include six cavernicolous species, no less than five of which are troglobitic. Of the troglobites, only two, *Niphargus fontanus* and *Niphargellus glenniei* occur with any frequency in caves and the latter is confined to Devon, where the former does not occur. The geographical and ecological distribution of the three *Niphargus* species and *Crangonyx subterraneus* is interesting and far from easy to interpret. It would seem that the odd-man-out is the widely-distributed *Niphargus aquilex,* which inhabits superficial groundwaters, such as river gravels (both in soft and hard waters), a habitat not apparently much frequented by the other three species. *Niphargus fontanus* appears to inhabit mesocaverns and caves in the unsaturated zone above the water table in calcareous rocks as well as phreatic mesocaverns and caves. *Niphargus kochianus* and *Crangonyx subterraneus* also occur in flooded mesocaverns and are frequently found in wells, but seldom appear in caves. The geographical ranges of all three species overlap in central southern England, as far west as the Mendip Hills, and *Crangonyx* and *N. fontanus* also occur in south-east England and South Wales. *N. kochianus* is absent from the latter areas, but occurs in East Anglia. Hazelton and Glennie (1953) have reported that *Crangonyx* kept in captivity

"was seen to crawl upright in the silt, and never swam or lay on its side ... quite different from the behaviour of *Niphargus* species".

What this signifies about its mode of life, I am not sure.

In Britain *Gammarus pulex* is commonly found in cave streams and less frequently in isolated seep-fed pools. One population which I have observed repeatedly over a fifteen year period inhabits a series of seepage-fed crystal-lined pools in a high oxbow, some 20 m above the cave stream of Swildon's Hole in the Mendips. All the members of the population are largely depigmented and most have chalky-white eyes. There is no doubt that this is a cavernicolous population. Cave populations of *G. pulex* in Yorkshire have been studied by Piearce and Cox (1977), who reported that individuals of a pale yellow-orange colour taken from Ingleborough Cave showed a greater sensitivity in hiding from light than did normally-pigmented specimens from Clapham Beck (which has its source in the cave). This was evident even after a two-week acclimatization period when the two sample groups were maintained under identical regimes of light-and-dark cycles. A student of Piearce's, Lori Farmer, subsequently found that "size for size the first antennae of unpigmented *Gammarus pulex* are relatively thinner than those of pigmented [specimens]" and Piearce also records that on examining the eyes of specimens from Ingleborough Cave under the microscope, "the spaces between the pigmented ommatidia seemed much larger in the cave *Gammarus* than in those from Fell Beck". Piearce wonders whether in *Gammarus pulex* we may have "a cave species in the early stages of its evolutionary adaptation to subterranean life, not yet a new species, or form perhaps, but on the way towards it."

St Patrick must have had it in for cavernicoles as well as snakes, for Ireland has only a single troglobitic amphipod, *Niphargus kochianus irlandicus* (as distinct from the British subspecies *N. k. kochianus*). It occurs in caves and groundwaters in Clare, Galway, Limerick, Tipperary, Offaly, Kildare and possibly Dublin and has been dredged from the depths of Lough Mask (in Co. Mayo), which is apparently fed by deep springs. Irish caves even lack *Gammarus pulex*, whose place is taken instead by *G. duebeni*, although it has yet to be established if that species can maintain permanent underground populations in the same way as its cousin across the water.

Symphyla

The Symphyla are essentially soil-dwellers and are usually reckoned to be of little interest to cave biologists. However, *Symphylella isabellae* (like a small, anaemic, scolopendromorph centipede) is not infrequently recorded from caves and is known to feed on organic detritus and should therefore be regarded as a possible cavernicole.

Diplopoda

Some 25 species of millipedes have been recorded from British and Irish caves, and a few are undoubtedly cavernicoles. Of these, the commonest and most widely distributed is *Polymicrodon polydesmoides*, a fawn-to-brown coloured species up to 2 cm long, with the typical flat-backed appearance of the Polydesmoidea. Turk (1967) has found that in underground populations of this species the number of ocelli (simple eyes which often occur together in clusters) varies from 0–23 as compared with a maximum of 30 or so in surface individuals, and he suggests that some populations may be on the way to becoming troglobites. The rare millipede *Brachychaetuma melanops*, a white, slender-bodied species, has an interesting distribution. It is found in patches along the warm southern coasts of Dorset, Devon and Cornwall, but also in

Fig. 4.28 Spotted snake millipede *Blaniulus guttulatus*, rolled up in defensive posture.

Otter Hole near Chepstow. The latter population may be a cavernicolous relict from a period of warmer climate when the species may have enjoyed a wider distribution.

At least three other species in Britain may also be viewed as cavernicoles: *Brachydesmus superus* is a small, pale polydesmid with a similar body-shape to *P. polydesmoides*, found in caves in South Wales, Derbyshire and Mendip. *Blaniulus guttulatus*, the spotted snake millipede, is a long, slender, white-bodied julid millipede species with an orange-red dot on the side of each segment, which occurs in caves mainly in south-west England. *Polydesmus angustus* is found rather more commonly in caves in continental Europe, but also occasionally in British caves, mainly in Yorkshire and South Wales. *Brachydesmus superus* seems to be the most frequent millipede in Irish caves. It has been recorded in Clare, Tipperary and Fermanagh.

Chilopoda

Centipedes are seldom found in British caves. Of nine species recorded to date, only *Lithobius duboscqui*, a small, cryptic and rather short-bodied species is considered to be a possible cavernicole.

Insecta

Current estimates suggest that around 95% of all animal species on earth may be insects. There are more than 20,000 species on the British and Irish lists, representing 25 orders, and several hundred of these have at some time or other been recorded in caves. Fortunately, for our purposes in this chapter, only three orders – the Collembola, Diptera and Coleoptera – are significantly endowed with cavernicoles.

The Collembola, or springtails, are a difficult group for the cave biologist. All are small, or minute, and the systematics of many genera are so confused that it is impossible in many cases to secure species identifications for material collected. Springtails are hygrophilic, often lack pigmentation, and feed on moulds, fungal hyphae, bacteria, other tiny invertebrates or organic detritus, making them ideally suited to life in the mesocaverns. A large number of species has been recorded in caves in situations which suggest that they are maintaining cavernicolous populations, and at least four of these are associated so closely with caves in Britain that they must be regarded as troglobites. They are *Onychiurus schoetti* and *O. dunarius* – slender, white, rather inactive rod-shaped creatures with short, club-shaped antennae; *Pseudosinella dobati* – an ivory-coloured, energetic entomobryid with a distinctive ruff of club-shaped hairs behind its head; and the rare *Pararrhopalites patritzii* – a small, white, round-bodied leaper which carries its long antennae like the claws of a scorpion, and frequents pool surfaces. A rather larger and commoner relative of the this last species, *Arrhopalites pygmaeus*, is considered to have a cave-limited distribution in parts of Germany and France, and is a common inhabitant of the deeper parts of British caves.

In Ireland, according to Mary Hazelton (1974), the Collembola most frequently recorded are a *Folsomia* species which occurs in the caves of Cork, Clare, Sligo and Fermanagh and *Anurida granaria*. Hazelton considers the *Folsomia*, together with *Schaefferia emucronata*, *S. willemi* and *Onychiurus schoetti* to be troglobites.

World-wide, many species within the large order of the Diptera, or true flies, are found in caves, but surprisingly few have been recognized as being troglobites, and none at all in Britain and Ireland.

Fig. 4.29 A newly-emerged adult fungus gnat *Speolepta leptogaster* dries its wings. Its pupal case hangs close by among the remains of the larval web.

Fig. 4.30 The larva of *Speolepta leptogaster* uses its moisture-beaded web as a scaffolding to cover a relatively large area of the cave wall in its search for food.

The Mycetophilidae, or fungus-gnats, include a number of cavernicoles, the most interesting being *Speolepta leptogaster* whose winged adults have occasionally been recorded in non-cave habitats, but whose larvae are known only from caves, with a wide distribution throughout Britain and Ireland. The adult is olive-brown in colour and is a typical mycetophilid shape – a bit like a miniature crane-fly, although with proportionally shorter legs. The long-bodied, translucent larva may grow to 14 mm long and specimens are occasionally recorded as having depigmented ocelli. It constructs and spends its entire growing life in a tangled scaffolding of threads spun across the surface of the rock wall. Some related tropical species (such as the famous New Zealand cave glow-worm, *Arachnocampa luminosa*) use similar webs to catch prey, but in *Speolepta*, its main function would seem to be to keep the animal clear of the wall. This might suggest that the principal habitat of *Speolepta* is in solutionally-opened joints of mesocavernous dimensions, where the web would allow the larva to avoid being washed away by meteoric flow down the walls of the cracks, following rain on the surface. In caves, the larvae change their position from time to time, spinning new webs as they go. They are often to be seen moving along the threads and while thus suspended, the head is slowly passed to and fro over the cave wall in a sweeping movement. This may be interpreted as feeding behaviour, and it is possible that the larvae graze on the microflora growing on the damp cave wall. Gut contents tend to be an amorphous sludge containing perhaps a few distinguishable fungal spores, which does not provide much help. Mature larvae may migrate to drier areas of wall or the cave ceiling where pupation takes place. The larvae shed their skin and pupate inside a temporary mucus tube.

Members of the genus *Sciara* (subgenus *Bradysia*) of the family Sciaridae are commonly reported in caves. These little gnats are very similar in shape and habits to Mycetophilidae, but the adults are dark-coloured, while the slender, translucent larvae have a black head-capsule (it is straw-yellow in *Speolepta*). For the present, the systematics of the group are in a mess, so that it is not possible to assign species names to those collected in caves, however a close search under stones, on walls or stalactites in certain areas will often disclose the threads or loose webs (commonly mis-attributed to the work of money spiders) made by the larvae, which appear to feed on decaying organic matter or moulds.

The trichocerid winter-gnat, *Trichocera maculipennis* is frequently found deep in caves and is much more common there than in surface habitats. If protein baits, such as rotting liver, are put down, it is not long before they become a wriggling mass of dirty-yellow, maggot-like trichocerid larvae, yet at other times the larvae are seldom recorded, and the normal behaviour of the species underground remains unknown.

Triphleba antricola, a member of the rather more advanced family Phoridae, or coffin flies, occurs widely in caves and is rare elsewhere, as is *Limosina racovitzai*, in the family Sphaeroceridae.

According to Hazelton (1974), a chironomid midge discovered by W.B. Thomas deep in the Doolin Cave system of Co. Clare turned out to be a species of *Tanytarsus (Rheotanytarsus)* not previously known to the British *(sic)* fauna. The larval cases were found clinging to pebbles in shallow, swiftly-flowing water between 160 m and 300 m upstream from Fisherstreet Pot, but it is not clear what ecological dependence this species may have on caves.

By far the largest order in the animal kingdom, the Coleoptera, or beetles, include a very large number of troglobitic forms in other countries, but sadly not in Britain and Ireland.

Perhaps the most interesting beetle found in British caves is the carabid *Trechus micros*. This is a small, orangey-brown coloured ground beetle which has been recorded from caves in Mendip, Derbyshire and Tipperary as well as in many parts of Europe. The blind larvae have rarely been found, and then only in caves, although the adults also occur in and may be primarily adapted to the mesocavernous spaces of the SUC and screes. In caves, the beetles seem to prefer damp, muddy situations, such as occur in the entrance series of Otter Hole near Chepstow and Manor Farm Swallet on Mendip, which both support substantial populations. In Ireland, two larger carabid beetles: *Trechus fulvus* and *T. obtusus* (in addition to *T. micros*), are also thought to maintain cavernicolous populations.

Among the Staphylinidae, or rove beetles, three species: *Quedius mesomelinus*, *Lesteva pubescens* and *Ochthephilus aureus* are widely distributed in British and European caves and are probably cavernicolous. A fourth species, *Aloconota subgrandis*, has, in Britain, only been found in Otter Hole, although there is no direct evidence that it breeds underground.

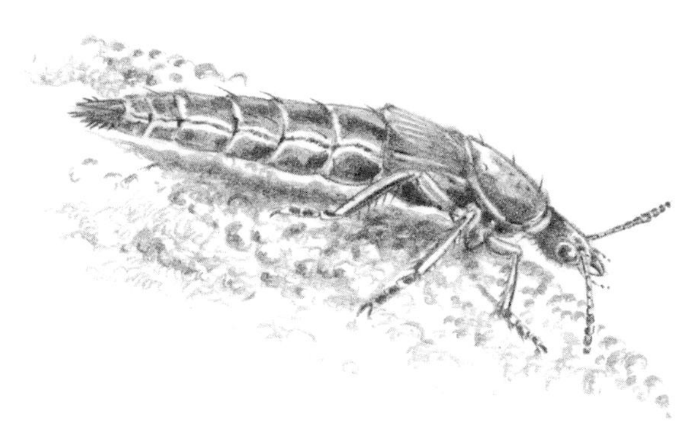

Fig. 4.31 The rove beetle *Quedius mesomelinus* is particularly widespread in European bat caves, often in association with guano deposits.

Most of the many species of water beetles recorded in our caves are probably of accidental occurrence, having been washed in by sinking streams, but four species belonging to the family Dytiscidae are probably cavernicolous. *Agabus guttatus* and *A. biguttatus* are found outside as well as within caves, but two smaller diving beetles, *Hydroporus ferrugineus* and *H. obsoletus* are known only from subterranean habitats in Britain.

As a postscript to this account, I should perhaps mention one other rather interesting 'cavernicole', if the term may be applied to a species which survives on the detritus of human activity in a man-made hole. In the course of a faunal survey of South Wales coal mines in the 1940s, the introduced American cockroach, *Periplaneta americana* was found in several pits. This is a supremely opportunistic species, which has an affinity for detritus and for dark, humid places. In warm, temperate or tropical countries *Periplaneta* has colonized a large number of caves in the wake of human disturbance, often decimating the native arthropod fauna. British caves are too cool and generally lack the food resources needed by the species, but deep coalmines seem to provide an acceptable habitat. The population inhabiting one particular pit in South Glamorgan was found to contain a white-eyed mutant form of *Periplaneta*, which was subsequently cultured and studied by Jeff Jefferson at University College, Cardiff. Breeding experiments showed that the mutant character behaves as if produced by a single, simple recessive gene, and a series of surveys of *Periplaneta* in the mine itself over 11 years, showed that the mutants comprised about 5% of the population, with no tendency towards any change in this proportion throughout the period.

Submariners

The submarine entrances to the Green Holes of the Hell complex open along the base of a submerged cliff at a depth of around 10 m below the low tide limit (chart datum) of the exposed Doolin Point in Co. Clare. Just to the south, at slightly greater depth, lies a second submerged system – the Reef complex, while to the north, the much longer Mermaid's Hole runs inland for almost one kilometre. Swept by strong tidal currents which make exploration extremely hazardous except in the calmest weather, the caves hold a unique fascination for cave divers. Swaying kelp forests (*Laminaria digitata* and *L. hyperborea*) surround several of the entrances, lending – according to diver John Adams – "an air of marine beauty unparalleled anywhere else in the British Isles." A couple of hundred metres north of Mermaid's Hole lies the largest of the intertidal Brown Holes. Similar, but less extensive submarine caves occur around Brixham in Devon, and close to Durness in northern Scotland.

At the time of writing, no detailed account of the biology of submarine caves in the British Isles has yet been published, so I am particularly grateful to those marine biologists, such as Mark Woombs, and other diver-naturalists who have shared their observations of the life to be found in such places.

The Green Holes, off Doolin, and the Brixham caves were first systematically documented by Peter Glanvill, who noted the abundance of encrusting and bottom-dwelling organisms in submarine caves when compared with the poor fauna of their freshwater counterparts. He was particularly struck by the abundance of forms normally found at much greater depth in the open sea.

The dimly lit entrance passages of the aptly-named Green Holes offer a sheltered habitat which encourages the proliferation of delicate encrusting organ-

Fig. 4.32 In the sheltered confines of Anemone Arcade, part of the Hell complex off Doolin Point, Co. Clare, larger Dahlia Anemones, *Urticina felina*, festoon the walls and Common Prawns, *Leander serratis*, grow to giant proportions, sustained by a plentiful food supply washed in by the surge.

isms such as tube worms, Cnidaria and sponges. By far the most common organism is the tube worm *Spirorbis borealis*, whose white limy tubes may cover up to 50% of the cave walls, especially near entrances. The keel worm *Pomatoceros triqueter* is much less abundant, accounting for a mere 1% cover. In sheltered tunnels, the mouths of worm tubes may be prolonged outwards to curl like slender villi into the lumen of the passage, giving the occupant a better chance to snatch life-sustaining particles of food from the passing currents.

Porifera of various sorts have been recorded by Mark Woombs. The commonest is a Breadcrumb Sponge, *Halichondria panacea*, which forms pancake-like patches, sometimes in unusually vivid colour-forms, covering up to 2% of the cave wall. A pure white form of the Elephant Ears sponge, *Pachymatisma johnstonia*, is locally common in the Hell Complex, in places covering up to 5% of the cave walls. Other notable species are the black sponge *Dercitus bucklandi*, which is typical of dark undersea caves, and the pit-boring *Cliona celata*.

The most beautiful cave inhabitants are delicate Cnidaria, such as the Devonshire Cup Coral, *Caryophyllia smithii*, the Jewel Anemone, *Corynactis viridis*, and the Oaten Pipes hydroid *Tubularia*. Devonshire Cup Coral is a solitary species more closely related to sea anemones than to the colonial reef-forming corals of the tropics. It is particularly common in the Brixham caves. The cup grows a centimetre or so high and wide and is attached to the cave wall by a short stalk. The radially-ribbed mouth of the cup, which may be white or pink, encloses the animal itself, which has about 50 delicate stinging tentacles with brown markings and white tips. The Jewel Anemone is similarly patterned in pink and white and forms spectacular close-packed colonies covering up to 10% of the cave walls in some Green Holes.

Urchin Cave, the largest of the intertidal Brown Holes on the coast of Co. Clare, is particularly well endowed with cup corals. It is possible to walk into the entrance passage of this cave at low tide, when the unretracted polyps of *Caryophyllia* may be seen dangling out of their cups. The same cave contains colonies of the shallow-water Beadlet Anemone, *Actinia equina*, and of *Sagartia elegans*. Chris Proctor has recorded a beautiful pale green form of *Sagartia* as abundant in the marine caves in the Brixham area, and Mark Woombs has noted very unusual bright orange and vivid turquoise colour forms of *Actinia* in the Doolin caves.

Anemone Arcade, in the Hell complex off Doolin Point, was named for the presence of spectacular Dahlia Anemones, *Urticina felina*, which grow up to 15 cm tall and are big enough to catch small fishes. Diver Rob Palmer reports seeing them trapping and swallowing pieces of Compass Jellyfishes, *Chrysaora hysoscella*, which had been swept into the cave in large numbers by a storm.

The colonial cnidarian *Alcyonium*, ghoulishly known as Dead Man's Fingers, is found in the entrances of some Brown Holes and the Reef Caves off Doolin Point. The colonies consist of a mass of polyps embedded in a finger-like body mass which is strengthened with calcareous fragments, giving it a flexible, rubbery feel. Like the hydroids and sponges with which they share the cave walls, *Caryophyllia, Alcyonium* and *Corynactis* feed mainly on plankton. A Sea Squirt (probably *Ascidiella* sp.), recorded in Chert Ledge Cave, is another sessile plankton-feeder, although much rarer in the Green Holes.

Glanvill has noted that the numbers of encrusting organisms is greatest in those passages which open to the sea at both ends, and which therefore combine a good current flow (bringing a daily supply of food) with a high degree of shelter from predators and storms.

Echinoderms are very common in our marine caves. Urchin Cave, mentioned above, is so-named after the large numbers of Common Sea Urchins, *Echinus esculentus*, which cluster inside the entrance at low tide, emerging to browse on the algal-covered rocks when the tide comes in. The darker passages of the Green Holes also contain large numbers of this species, which may reach 16 cm diameter and which fully retain their striking crimson and purple colours despite the absence of light. The urchins seem equally at home on the roof, walls or floor of the cave, where their long, slender tube-feet stream out in the current, fulfilling the role of gills through which the animals excrete and breathe. In Brittlestar Boulevard, in the Hell complex, multitudes of Brittle Stars, *Ophiothrix fragilis*, crawl in writhing heaps over the boulder floor, waving their arms about in search of food. As crevice-loving detritivores, Brittle Stars are perfectly preadapted for cavernicolous life. They appear to congregate in this particular tunnel because this is the point at which detritus, swept in by the tidal current, is dropped from suspension. In Harbour Hole (in the Reef complex off Doolin Point), Brittle Stars have been observed lurking in holes in the cave wall with their arms poking out and waving about in the current.

Rosy Feather Stars, *Antedon bifida*, are commonly seen clinging to the walls of Harbour Hole, one of the Doolin Reef Caves. These primitive echinoderms, related to the ancient crinoids frequently found as fossils on the walls of caves, are filter-feeding detritivores which use their long, pinnately-feathered arms to trap edible particles swilling about in the water. Feather Stars are delicate organisms, generally found in very sheltered conditions well below the tidal limit. Short, clawed stalks below the disc from which the arms radiate, allow the animals to cling firmly to the substratum, and this is how they are usually seen in caves. If the need arises, they can swim most gracefully, beating five arms upwards and five downwards alternately. Another detritivorous echinoderm, the Cotton Spinner *Holothuria forskali* (a dark-backed sea cucumber up to 20 cm long) also inhabits Green Holes. When disturbed, it can confuse or entangle an attacker by shooting out a mass of sticky, cotton-like threads from its anus.

The Common Starfish *Asterias rubens* and the Spiny Starfish *Marthasterias glacialis* are present in the Hell complex in such numbers that marine biologist Mark Woombs considers it probable that they must live in these habitats permanently. Although starfishes are generally thought of as predators, feeding on bivalve molluscs (such as Mussels) which are not found underground, they are also opportunistic scavengers, and Glanvill reports seeing them scavenging on dead fish remains in the caves.

Most of our larger inshore Crustacea are long-lived, rather cumbersome, nocturnal scavengers, adapted to rocky coastlines which offer shelter from fleeter enemies which relish their tender flesh. The huge-clawed Common Lobster, *Homarus vulgaris*, (which may weigh as much as five kilograms) and its smaller-clawed cousin the Crawfish, or Spiny Lobster, *Palinurus vulgaris*, are commonly seen at the entrances to Green Holes, and occasionally well into the dark zone. Both species are greatly in demand by gourmets (the Crawfish is the French 'Langouste') and are now in decline over many parts of their range owing to the pressures of over-fishing. 'Spider Crab Crawl', in the Hell complex, is so named after the Spiny Spider Crabs, *Maia squinado*, which, together with Edible Crabs, *Cancer pagurus*, occasionally enter this and other Green and Brown Holes.

Galathea strigosa, a Squat Lobster with striking scarlet and royal-blue colours, is common throughout the Green Holes and in the Brixham caves. Squat Lobsters are in some respects pre-adapted for cavernicolous life, and an eyeless troglobitic species has evolved in submarine lava tubes in the Canary Islands. The ubiquitous Common Prawn, *Leander serratus*, is obviously very much at home underground. Specimens up to 12 cm long (twice the normal size) are frequently seen scavenging on the walls of the Irish Green Holes. They must have few natural enemies in the caves, for Woombs reports that they are quite fearless and will happily sit on a divers arm, in marked contrast to their usual skittishness in the world outside.

Molluscs are not well represented in our submarine caves, perhaps because there is little for them to eat. The only common species is the Thick-lipped

Dog Whelk, *Nassarius incrassatus*, although Woombs reports seeing a Mussel, and small numbers of Cowries and several species of Sea-slugs in the Doolin caves.

A number of fishes are reported from our marine caves. The most common species seems to be the Conger Eel, *Conger conger*, a well-known frequenter of dark holes and wrecks, which may reach over four metres in length and has a reputation for belligerence when disturbed in its lair. On a smaller scale, the Leopard-spotted Goby, *Thorogobius ephippiatus*, a renowned cave or crevice-dwelling species, was recorded by Woombs in the Hell complex. Woombs also reports seeing many Wrasse in the entrances to the Doolin caves, including the Cuckoo Wrasse (*Labrus mixtus*), Ballan Wrasse (*L. bergylta*), Rock Cook (*Centrolabrus coletus*) and Gold Sinny (*C. rupestris*), and also a Pollack (*Pollachius pollachius*). Glanvill has seen a dogfish of some sort apparently asleep on the floor of Harbour Hole, part of the Reef complex just south of Doolin Point. Nearby Chert Ledge Cave harboured a Three-bearded Rockling, *Gaidropsaurus vulgaris*, and a Tompot Blenny, *Parablennius gattorugine*. Palmer photographed a Topknot, *Zeugopterus punctatus*, on the floor of a passage in the Hell complex, and this species apparently is also found in marine caves in Devon.

5

Cave Communities

In the last chapter we considered the biota of caves in Britain and Ireland taxon by taxon, according to the individual species' degrees of dependency on a particular zone of the cave. In this chapter, we will consider a portion of this same fauna from a different perspective – examining how commonly co-occurring species interact with each other within various recognizeable habitats in the dark zone of the cave. In other words, this chapter is about how individual cave communities, or associations, work. In this context, a 'community' is defined as a group of organisms which spend all their lives together in one place, while the members of an 'association' do not necessarily all come together in the same place at the same time, although each spends a certain time in association with a particular place or resource common to all other members.

The wall association

The walls and ceiling of the deep threshold, and dark, variable-temperature zone of the cave provide relatively predictable gradients in microclimate. Visiting insects, by periodically shifting their positions, are thus able to maintain themselves in an optimal environment during the period of their stay. Their presence year after year in the same parts of the cave provides a dependable resource which is exploited by a number of predatory and parasitic cavernicoles, including fungi, snails and arachnids.

The visiting insects which form the basis of this community are the Herald Moth (*Scoliopteryx libatrix*), Tissue Moth (*Triphosa dubitata*), three Diptera (*Culex pipiens, Limonia nubeculosa* and *Heleomyza serrata*) and the caddis fly *Stenophylax permistus*. Other insects also occur from time to time, but the above species are by far the most regular visitors.

Apart from entomophagous fungi, such as *Hirsutella* and *Beauveria*, the major predators of insect members of the wall association are the two spiders, *Meta merianae* and *M. menardi*. The former lives in the threshold, within the area reached by daylight, while the latter prefers the deep threshold at, or just beyond the limit of light penetration. Both species produce sticky orb webs, but these are rather small affairs – disproportionately so in the case of the large *M. menardi*, which produces a web more in keeping with a spider half its size. Surprisingly, the web is oriented not across the cave passage, where it might be expected to capture the maximum amount of prey, but parallel with the cave wall. The reason for this becomes obvious if an examination is made of the rubbish which collects below the web. This is found to consist mainly of sections of the exoskeletons of woodlice, millipedes and beetles, creatures more typical of mesocavernous cracks than of larger cave passages, and indeed the spiders would not make a particularly good living if they relied solely on the very sporadic inputs of flying insects of the species previously listed. As it is, they show adaptations to a poor food supply in the form of reduced activity coupled with low basal metabolic rate.

Fig. 5.1 A cave snail *Oxychilus cellarius* attempts to eat a hibernating Herald Moth *Scoliopteryx libatrix*. The snail secretes chitinase enzymes which digest insect cuticle, but such attacks do not always succeed – the torpid moth may rouse itself in time to escape.

In a comparison with the common orb-web spider *Araneus diadematus*, specimens of *M. menardi*, matched for size and maintained under similar environmental conditions were found to consume oxygen at one eighth the rate of their surface cousins.

The egg sac of *Meta menardi* is an impressively large (25 mm long by 20 mm wide), balloon-shaped structure spun of delicate, pure white silk, typically supended from the roof of the cave by a pedicle about 25 mm long and guyed in position by fine silken lines. It contains from 400–500 bright yellow, spherical eggs. The spiderlings may disappear into mesocavern cracks soon after hatching. The species is reputed to be unusually tolerant of high carbon dioxide levels (up to 15%) which may be an adaptation for a mesocavernous life during the early stages of the life cycle.

An unexpected predator of the wall association is the snail *Oxychilus cellarius*, whose feeding behaviour in the caves of Belgium has been described by Georges Thines and Raymond Tercafs. Zonitid snails, such as the cavernicolous *Oxychilus* species and *Vitrea crystallina*, secrete chitinase enzymes which digest the exoskeletons of insects. Chitinases are surprisingly rare in nature, so that most insectivorous animals such as bats, shrews, birds and spiders, are able to digest only the soft parts of their insect prey, while the hard parts are rejected or ejected. In caves, insect cuticle is a relatively common resource, present in the fungus-ravaged remains of *Heleomyza serrata* flies on the cave walls, in bat droppings on the floor, and in the frass which accumulates below spiders' webs. *Oxychilus cellarius* routinely scavenges on such material, but according to Thines and Tercafs, it will also on occasion attack living insects and especially moths.

The Herald Moth, *Scoliopteryx libatrix*, is particularly at risk as it passes much of its time in a deep torpor. Apparently, the snail has been observed to crawl up on to sleeping moths and eat portions of their wings. In another instance an attack was made on the abdomen of the moth. The snail rasped away the dorsal cuticle, insinuated its head into the abdominal cavity and ate away some of the underlying soft tissue. The attacks rarely last long, and are not always immediately fatal, as the moth is sometimes able to rouse itself and struggle free. Interestingly, in fatal attacks, the wounds inflicted by the snail do not grow mouldy after death, suggesting to the original observers that *Oxychilus* might have been secreting some form of antibiotic in its saliva. However, later laboratory tests showed that the tissues attacked by the snail were heavily infested with bacterial colonies which appeared to be excluding the more usual fungal decomposers. Another interesting observation made by Thines and Tercafs is that whereas outside caves *Oxychilus* species normally seal up the entrance to their shell during winter and remain in hibernation until the onset of spring, in caves they remain active all year round.

Millipedes, and particularly the large craspedosomid *Polymicrodon polydesmoides* are regularly seen on cave walls where they may scavenge on carcasses of visiting insects or their fungal decomposers. Fungal hyphae associated with dead insects, the droppings of millipedes, and the frass found beneath webs of the large *Meta* spiders may attract various Collembola, which may feed on any of these potential energy sources or on the microfungi which grow on the wastes of the spiders and millipedes. The Collembola in turn may fall prey to spiders such as *Nesticus cellulanus*, which in European caves is considered an integral part of the wall association.

Terrestrial mud bank community

Organically-rich mud provides one of the principal sources of food for cave communities. Bits of rotting leaves and other such detritus may be eaten directly by worms, millipedes or isopods, but most mud-bank cavernicoles prefer to browse on the decomposing micro-organisms which grow on the surface of the silt, or prey on the browsers themselves.

Perhaps the most common arthropods found on mud banks are springtails belonging to the families Isotomidae (several *Folsomia* and *Isotoma* species), Hypogastruridae (*Schaefferia emucronata* group), Tomoceridae (particularly *Tomocerus minor*) and Entomobryidae (particularly *Heteromurus nitidus*).

There have been no studies of the feeding behaviour of Collembola in British caves, but work by European and American cave biologists has established a strong correlation between the abundance of most cavernicolous Collembola and the presence of microfungi in cave sediments. In France, *T. minor* is known to feed principally on microfungi, and it may be that most other British cavernicolous Collembola do so too, although *Folsomia candida* seems to be

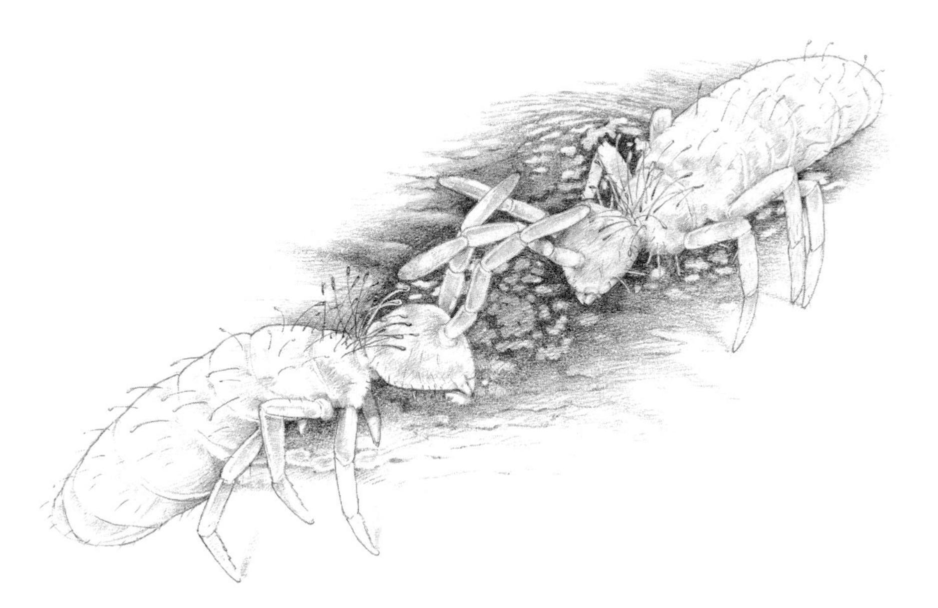

Fig. 5.2 Battling cave springtails, *Pseudosinella dobati*, dispute territorial rights to pastures of microfungi, which they browse like miniature cattle.

drawn to smelly meat baits, particularly on wet substrates, suggesting a preference for bacterial rather than fungal food.

The principal bacterial feeders in cave sediments are likely to be nematodes and oligochaete worms. Not a great deal is known about cavernicolous nematodes, but cave-dwelling worms have been studied in some detail in Ingleborough Cave in Yorkshire by Trevor Piearce. The two most common species in the dark part of the cave were *Allolobophora chlorotica* and *A. rosea*. Piearce recorded that the sediment banks in one passage in Ingleborough Cave were completely covered with worm casts, although neither of the above species normally produces casts on top of the soil. I have noted a similar phenomenon in Otter Hole, where the principal species responsible is again *A. chlorotica*. The normal habitat of these worms is soil with a good crumb structure, where the numerous cavities created by plant roots provide ample space to deposit egested material below ground. In the cave, the sediment banks are far more compact and the worms are therefore obliged to cast on the surface.

The Ingleborough Cave worms were frequently seen crawling on the surface of sediment banks and even on the roof of the cave, in marked contrast to their behaviour outside of caves, where they are seldom seen at the surface. Worms of other species, such as *Dendrobaena rubida* and *Eiseniella tetraedra*, are also found in the open like this in the 'deep cave', probably because the lack of predators, the darkness and the 100% relative humidity of their surroundings liberates them from the need to remain beneath the soil surface and so encourages more active foraging. The organic detritus on which the cave worms depend for food is relatively scarce and tends to be deposited at the strand line of the winter and spring floods. This may be on the mud banks where the worms habitually live, or half way up the cave walls or on the roof.

Millipedes such as *Polymicrodon polydesmoides*, *Brachydesmus superus* or *Blaniulus guttulatus* are well adapted to exploit particulate detritus stranded by floods, as they can move rapidly over the substrate and are adept at climbing up vertical surfaces. Cave millipedes often occupy broad detritivorous niches, feeding on virtually anything organic, from dead animals, to bacterial silt, rotting leaves and microfungi. *Polymicrodon* seems particularly tolerant to inundation by floods and is quickly on the scene as the waters recede. Its droppings are often so common on sediment banks and beneath vertical cracks in the cave ceiling that it is difficult to account for the relatively low numbers of sightings of the species in caves. I would guess, in the absence of evidence either way, that it must be very common in mesocavernous spaces below the soil.

While the larger lumbricid worms may have few predators below ground, small enchytraeid pot worms and tubificids may fall prey to predatory beetles, such as *Trechus micros* or *Lesteva pubescens*. I have watched the former species as it worked over a tidal mud bank in Otter Hole Cave, pocking the surface with dozens of tiny pits as it thrust its head repeatedly into the silt, presumably in search of such prey. These and other cave beetles may also take Collembola, although I know of no confirmed observations.

Spiders of the genus *Porrhomma* are clearly well adapted to feed on springtails, but are not common on mud banks. Pseudoscorpions, which are important predators of Collembola in leaf-litter, have hardly ever been recorded from British caves, although a number of highly cave-evolved species are found in Europe and elsewhere. Finally, predatory mites, such as *Rhagidia spelaea*, are often to be seen running over mud banks at great speed but are seldom observed actually doing anything of biological significance. No doubt in the normal darkness of the cave, all kinds of interesting and unrecorded behaviour continues right up to the instant that the underground naturalist floods the scene with light, when the whole cave community, such as it is, promptly freezes to the spot, dives for cover, or runs off in a panic.

'Batellites'

British caves lack the large, year-round bat roosts so common in the tropics. Nevertheless, our caves contain a few species of invertebrates whose lives revolve around the presence of bats. Not all of these live among or feed on bat droppings, so that I prefer to categorize the typical cavernicolous fauna found in bat caves as a 'batellite', rather than a 'guanobious' community.

Perhaps the most familiar batellite found in caves is the tick *Ixodes vespertilionis*. Females of the species are blood-sucking parasites, while the males, which are thought to be non-feeding, are frequently abundant on the walls of bat caves and are seldom found outside. If the bats colonize a new cave, it soon becomes infested with ticks and so this essentially cavernicolous species has become distributed over a wide geographical area.

The carcasses of bats which die in the cave provide a rich food source for any species able to take advantage of the bonanza. In the world outside the cave, the odour of death may travel great distances on the wind, so that winged carrion-specialists can move from carcass to carcass in search of food for themselves, or in the case of insects, for their offspring. Carrion-feeding flies and beetles lay many eggs which hatch and grow as quickly as possible in order to pre-empt as much as possible of the available food supply. They invest all their energetic and reproductive resources in the search for the big payoff – a strategy which might be termed "putting all your eggs in one breadbasket". In British bat caves, carcasses are available seasonally and in small numbers, and so represent a particularly uncertain food supply. The trichocerid fly *Trichocera maculipennis* is one of very few cavernicoles which appears specialized to exploit this resource; within a week or so of a bat carcass appearing on the cave floor, it may have become a wriggling mass of *Trichocera* maggots, whose digestive enzymes rapidly turn the flesh into a semi-fluid goo, and the species is also associated with deposits of bat guano. Phorid flies, such as *Triphleba antricola* and *Phora* spp. may also utilize bat carcasses, but I know of no firm records.

Bat droppings are a far more dependable food resource, attracting a wide range of invertebrates, from relatively large snails and millipedes, to tiny mites and springtails. Zonitid snails, such as *Oxychilus cellarius, O. lucidus* and *O. draparnaldi*, utilize the fresh guano directly, as they are able to secrete an enzyme, chitinase, which allows them to digest the insect cuticle of which the droppings are chiefly composed. Millipedes such as *Polymicrodon polydesmoides* and *Blaniulus guttulatus* also seem to eat fresh guano, although it is not known precisely what they get out of it. Uneaten droppings are rapidly invaded by microfungi and moulds, and these in turn attract mycetophilid and sciarid flies, such as *Macrocera stigmata* and *Bradysia* spp. and Collembola, including *Onychiurus* spp. and *Schaefferia emucronata*. A number of mites (e.g. *Eugamasus loricatus* and nymphs of *Parasitus coleoptraterum* and *Lasiosius muricatus*) are also found around bat guano, but it is not clear whether they are associated with the droppings themselves or the bats which produced them.

Finally, predators are drawn to the guano by the presence of potential prey. The rove beetle *Quedius mesomelinus* is frequently reported from cave guano deposits, where it may feed on dipteran larvae or perhaps Collembola.

From this very brief account it may be deduced that not a great deal has been written about guano-associated faunas in British caves, and this would seem a fertile subject for future investigation.

Fig. 5.3 The bat tick, *Ixodes vespertilionis*, is widespread in bat caves across Europe, Africa, Asia and Australia.

Pool-surface associations

Pools fed by bedding plane seeps and dripping calcite formations may provide small oases of life within otherwise sterile fossil cave passages. The water they contain has come from the land surface above and will have brought with it organic material leached from the soil. Sometimes, particularly following heavy rain, the flow of water may have flushed small organisms from the mesocavernous cracks along its route, depositing them in the cave. Some, such as copepods, ostracods, *Bathynella stammeri*, or *Proasellus cavaticus* will be found in the pool itself, but others, such as Collembola and mites may accumulate on the surface film, together with a thinly-scattered flotsam of waxy or greasy organic detritus.

Collembola remain on the surface of pools because their bodies are water repellent. In most species the tegument is covered with a pattern of thickened ridges covered in a water-proof wax, separated by thin-walled troughs where the tegument is permeable to small molecules such as oxygen, carbon dioxide and water vapour. Most Collembola do not have a tracheal system and rely on gaseous exchange across these thin-tegument zones. A side effect of this system is that water vapour is easily lost in dry atmospheres. Troglobitic Collembola are unable to autoregulate transpiration and are therefore confined to sheltered, humid habitats, and are strongly anemophobic.

Troglobitic springtails have formed the subject of a remarkable study by Kenneth Christiansen. He has been able to distinguish between a series of 'cave-independent' characters – such as head chaetotaxy and labial papillae, which, though useful in determining phyletic position, show no consistent changes between facultative and obligate cavernicoles – and a separate series of 'cave-dependent' characters – such as the shape of elements of the claw complex and the size of the third antennal segment organ – which differ consistently between these two ecological groups. Highly modified cave species and those found deep in the soil, such as the Onychiuridae, all have an enlarged third antennal segment organ and this is thought to contribute to their extreme sensitivity to changes in humidity under experimental conditions.

Variations in claw structure are clearly related to the ability of Collembola to walk on wet, smooth surfaces, wet clay and water. The short claws of typical leaf-litter species, while ideally designed to grip on solid surfaces, are unable to penetrate the water meniscus and are consequently of little use on wet surfaces. In a remarkable illustration of convergent evolution, Christiansen has found that troglobitic Collembola are all equipped with long, slender, dagger-like claws, apparently designed to pierce the meniscus of the water surface in order to grip on to any solid material immediately below. Short-clawed Collembola are unable to climb the steeply-banked edge of the meniscus surrounding the cave pool. Each time they try, they simply slide back helplessly and their only chance of escape is to leap repeatedly in the air, until they fall by accident on to solid ground. However, the area surrounding the pool is seldom dry. More often it consists of waterlogged silt, which must present them with similar problems to those faced by a skiier on an ice slope set with boulders. Cave-evolved Collembola, on the other hand, are able to walk with ease on such surfaces. Their claws pass right through the thin curved water films which pose such a problem to their unspecialized cousins, so that they are able to clamber out of the pool at will. On our earlier analogy, the troglobites are like mountaineers equipped with crampons and ice-axes.

Christiansen considers that cave pools trap poorly-shod Collembola. Their starved carcasses, or the fungal decomposers which grow on them, no doubt contribute to the food resource available for cave-evolved species, which are drawn to the scene like hyaenas to a drought-stricken water hole. The species most at home on cave pools are *Onychiurus schoetti* and *Arrhopalites pygmaeus*, both of which are very widespread and successful cavernicoles in Britain. In a brief study of the Collembola present on 48 pools in Ogof Ffynnon Ddu (South Wales), I found a significant negative correlation between the occurrence of these two species on pool-surfaces, despite their similar distributions within the cave system as a whole, suggesting that they might interact competitively. Eight other species of Collembola were present on the pool surfaces in OFD. Two of these, *Schaefferia lindbergi* and *Folsomia* sp., were found only near the entrance and in areas where the fauna and microclimate suggested that the surface lay close overhead. They were strongly associated together on pool surfaces, which might suggest that they had been passively flushed into the pools from a shared habitat in the SUC which lies immediately above the cave at this point. Interestingly, the common troglobitic springtail *Pseudosinella dobati*, which showed a near-identical cave distribution to *Schaefferia* and *Folsomia*, was not found on pool surfaces at all.

Further away from the entrance, in portions of the cave which are separated from the land surface by a greater thickness of rock, a strong positive correlation was found between *Arrhopalites pygmaeus* and *Isotoma notabilis* on pools. It may be that these two species co-inhabit the deeper mesocavernous cracks

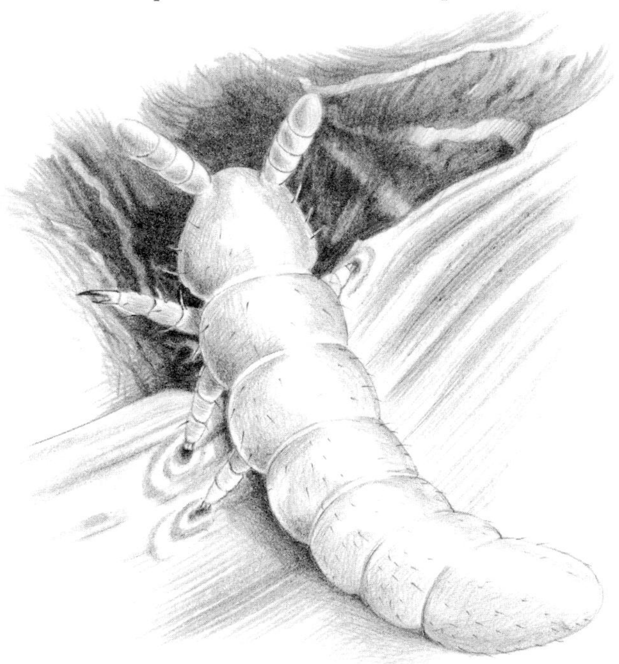

Fig. 5.4 *Onychiurus schoetti* climbing up a steep water meniscus, aided by its dagger-like claws.

between the SUC and the cave, or their principal habitat may be in the cave itself, in which case they presumably do not interact aggressively, as *Arrhopalites* and *Onychiurus* would seem to do.

The most interesting pool-frequenting springtail found during the course of the study was a single large *Onychiurus* apparently belonging to a previously unknown species in the *armatus* group.

Cave pools are frequently covered with a thin greasy-looking 'skin', perhaps derived from the decomposition of the carcasses of trapped insects. This organic layer may support a network of fungal hyphae, sometimes beaded with tiny sporangia carried on short vertical stalks. The large neanurid collembolan *Anurida granaria*, and hypogastrurids of the genus *Schaefferia* (particularly *S. lindbergi* and *emucronata*) seem particularly to favour 'eutrophic' pools of this sort, although they are by no means the only species to do so. Pools in stream passages may accumulate considerable numbers of dead caddis flies, providing a banquet for large gatherings of springtails of several species, some of which eventually add their own remains to the accumulating slick. Organically-rich pool surfaces are also frequently inhabited by fast-running predatory mites such as *Rhagidia spelaea* and *R. longipes*. These may attack and eat young springtails and adults of small species such as *Megalothorax minimus*, but presumably pose little danger to the larger well-defended adult onychiurids or hairy hypogastrurids. However, the tranquil water surface may harbour the odd nasty surprise. Cavernicolous flatworms, and in particular *Phagocata vitta*, are occasionally to be seen gliding about like patrolling sharks on the underside of the meniscus of cave pools. They scavenge on the floating bodies of dead arthropods and are reported also to take live prey on the water surface.

From the tentative nature of much of the discussion presented in this section, the reader will have gathered that the natural history of cave pool-surfaces is still poorly understood. It is an interesting area of cave biology which would repay future study.

Freshwater stream communities

Cave streams come in two varieties: those fed by runoff from non-limestone catchments, gulped down at swallets (allogenic waters) and those fed by rainwater percolating through soils and cracks in the limestone (autogenic waters).

Compared with swallet-fed streams, percolation waters in general tend to contain less organic material, are less subject to violent flood pulses (because large numbers of small inputs take varying amounts of time to conduct the extra flow after rainfall) and are subject to less fluctuation in pH and temperature. In the caves of southern England and South Wales, the most conspicuous inhabitants of autogenic streams and seeps are the isopod *Proasellus cavaticus* and the amphipod *Niphargus fontanus*. Both species are fairly common, although *Proasellus* seems to favour rather shallower water (stone-floored riffles), and perhaps organically richer conditions than *Niphargus*, and the latter is more typically found in silt-floored pools and in the saturated zone beneath the water table.

Proasellus may reach quite high densities in shallow, rocky streams where the substrate is stained with organic deposits or harbours flecks of rotted plant material. David Culver, an American zoologist who has made a detailed study of the ecology of aquatic cavernicoles states that "In stone-bottomed cave streams most of the food is plant detritus. Micro-organisms are present, but in very low numbers." (*Cave Life*, p 15. Harvard University Press, 1982). This observation suggests that Crustacea such as *Proasellus* which are characteristic of this habitat must surely be adapted to feed on plant detritus as a direct energy source. The gut contents of individuals found in such places are often dark coloured, but it is difficult to make out any recognizeable structures in the mush.

Proasellus are also frequently found crawling about in thin water films on flowstone and *gour-pool* terraces, particularly where 'water fungus' is present as a slimy layer on the calcite surface. Examination of the gut contents of *Proasellus* specimens collected in such situations shows a loose mass of rather fluffy-looking material consisting mainly of chlamydo-bacteria and other micro-organisms. Scrapings of the 'water fungus' show a similar composition and it seems reasonable to assume that the latter forms a major food item of the isopod. The floccular part of the 'fungus' may result from the physical aggregation of dissolved organic matter, as happens in the sea, or it might be actively precipitated in some way by the chlamydo-bacteria.

Niphargus in their silt-floored pools could be feeding on different micro-organisms. Culver (ibid) states that "In slow-moving streams ... with mud bottoms, and in drip and seep pools isolated from the main stream, microfungi are more common [than in stone-floored streams] and are correlated with the abundance of macroscopic invertebrates." According to Jefferson, the gut contents of *Niphargus* collected from such pools in South Wales usually contain a little silt, but also larger quantities of floccular material and micro-organisms similar to those found in the guts of *Proasellus*. Silt is known to play an essential part in the nutrition of some European species of *Niphargus*, apparently owing to the presence of vitamins and essential amino acids produced by bacteria living in it. There seems, however, to be no reason why organic matter in the silt, together with the microfungi which grow on it, should not also be utilized directly as a source of energy, and all the pool silts which have been examined have been found to contain both. Jefferson reported that silt in seep-fed pools in Ogof Ffynnon Ddu contained between 2.3 and 3.3% dry weight of organic material.

Niphargus is omnivorous and some species seem unable to develop to maturity unless their diet includes some animal matter. In captivity, *N. fontanus* will certainly prey on the smaller and more delicate *Proasellus* and there seems little reason to doubt that it will do so in caves. However, the isopods' flattened body and fleetness of foot must give them a good chance of escaping the larger and more unwieldy *Niphargus* in most circumstances. Perhaps a more dangerous threat is presented by cavernicolous flatworms such as *Dendrocoelum lacteum*. Predatory flatworms lay down a trail of mucus as they glide over the substrate, much as slugs and snails do. The mucus is sticky and acts as a trap for unwary Crustacea and as a chemical route-marker for the flatworm. Any passing *Proasellus* which is unlucky enough to get its feet mired in the trap is likely to be attacked and eaten by the returning flatworm.

Niphargus can sometimes be seen digging into the loosely-consolidated silt of cave pools in a way which suggests that they may be searching for food. Such substrates are known to support substantial populations of Protozoa, nematodes and enchytraeid 'pot-worms', and these may perhaps figure in the diet of *Niphargus*.

In the unsaturated zone, the population density of *Niphargus* is generally very low. During a faunal study in Ogof Ffynon Ddu II in 1978, I made a

Fig. 5.5 The predatory flatworm *Dendroselum lacteum* lays down a slime trail to trap food. A cave isopod *Proasellus cavaticus* unable to disentangle its feet from the snare, faces imminent death.

point of recording all the *Niphargus* seen during a score of visits to the cave, each lasting about four hours. On early visits, I generally saw one or two specimens. By the end of the study I had a fair idea of the best places to look, but still seldom recorded more than half a dozen sightings per visit – giving an average population density of about 2–3 individuals per kilometre of cave passage. This contrasts markedly with reports by cave divers of "hundreds" of the animals seen together in a short phreatic section of the main streamway, appropriately known as Niphargus Niche. Cave divers commonly report seeing 'white shrimps' during dives into resurgence caves, yet *Niphargus* species seldom appear in the springs themselves which is surprising, given that they are entirely eyeless and so might scarcely be expected to know when they had reached the limits of the cave. In fact, *Niphargus* is able to distinguish between light and darkness. According to Ginet, who has made a detailed study of the European species, the light receptors are in the antennae.

A characteristic of European species of *Niphargus*, which has not yet been reported in our caves, is their ability to survive droughts by burying themselves in chambers dug into the silt floor of pools as they dry out during the summer months. According to Ginet, the amphipods may survive out of water for more than ten months in this way, providing they are surrounded by a water-saturated atmosphere. I have followed the tracks of *Niphargus* across a muddy cave floor and in and out of shallow pools for up to a dozen metres in Ogof Ffynnon Ddu, and have seen shorter trails in a number of other British caves, so it would seem that *N. fontanus* (almost certainly the species responsible) is also to some extent amphibious.

The sediments in cave pools and streams frequently harbour small enchytraeid worms which may form an important prey item for *Niphargus* and for cavernicolous diving beetles and their larvae. The commonest of these, *Agabus guttatus*, favours shallow gently-flowing cave streams. Following heavy rain, its larvae sometimes turn up in seep-fed pools, suggesting that it may also inhabit mesocavernous spaces.

Allogenic streams often contain inwashed surface-stream species, such as *Gammarus pulex* (*G. duebeni* in Ireland) and various caddis larvae, particularly *Plectrocnemia* spp. which spin extraordinary nets beneath rocks to trap passing detritus. Perhaps the competition presented by these hungry invaders is too much for the true cavernicoles, or perhaps the physical and chemical regimes are intolerable to them – either way, it is rare to find either *Niphargus* or *Proasellus* in allogenic streams, at least in the unsaturated zone.

The possibility was raised in an earlier chapter that *Gammarus pulex* may be able to maintain permanent populations underground, although there is little doubt that most individuals seen in allogenic cave streams are first-generation immigrants. I know of no established *Gammarus* populations co-occuring with any species of *Niphargus*, which is interesting, as the former species not only survives, but appears to be able to breed successfully in microhabitats frequented by *Proasellus*. It would seem that the two amphipod genera are clearly adapted to very different lifestyles. The whole biology of *Niphargus* is centered on the need to economize energy (K-selection), while *Gammarus* species are adapted to effectively exploit abundant resources (r-selection). The two strategies dictate substantial differences in fecundity, metabolic rate, longevity and population dynamics between their respective adherents. So, we find that *Niphargus* produces far fewer and far larger eggs than *Gammarus* and incubates them for far longer (2.5–3 months compared with 2–3 weeks). The basal metabolic rates of typical members of the two genera, measured in terms of oxygen consumption at rest, differ by almost an order of magnitude. *Niphargus* species have a life expectancy of six years or more, while *Gammarus* species live for about two years. All of this adds up to the fact that where food is plentiful, the energetic and fast-breeding *Gammarus* will inevitably outcompete the more sedentary *Niphargus*, while the latter come into their own in conditions of poor and intermittent food supply. It would be interesting to know whether there is any trend among cave-breeding *Gammarus* towards the production of fewer and larger eggs, or towards a lessening of their basal metabolic rate.

6

Caves Through the Pleistocene

The age of ice

The geological period that includes most of the Ice Ages is known as the Quaternary. It began 1.6 million years ago and we are still in it. Thirty years ago, it was believed that there had been only four or five major glacial advances in this period, but it is now thought there may have been as many as seventeen cold-warm cycles, with an average duration of only 100,000 years. Evidence for these comes from measurements of the relative proportions of two isotopes of oxygen (O^{16} and O^{18}) incorporated into the skeletons of Foraminifera preserved in deep marine sediments. The lighter isotope evaporates more readily than the heavier, and so predominates in rainwater and glacial ice. During times of glacial advance, the ice sheets 'locked up' a significant proportion of the world supply of H_2O^{16}, increasing the proportion of O^{18} in the seas. Foraminifera incorporate sea- water oxygen into their shells, so that measurement of the relative proportions of the two oxygen isotopes in a core taken through a deposit of 'foraminiferal ooze' gives a fair indication of the prevailing climate at the time of deposition. The record so obtained can be tied in with more easily dated material, such as bone deposits in caves, to give a very good picture of climatic and ecological changes through the ages.

We are now in an interglacial, the Flandrian, which began about 10,000 years ago. The pattern of the last 1.6 million years suggests that interglacials are relatively abnormal periods, seldom lasting more than 10–15,000 years. In the longer term, there is no reason to doubt that, barring the effects of our own activities (in particular our enrichment of the atmosphere with 'greenhouse' gases, leading to an overall warming of the global climate), we should soon be plunged back into the 'normal' conditions of an ice age. Even within historically recent times the British Isles experienced a 'Little Ice Age' from about 1500–1850 AD, which elsewhere in Europe caused crop failures, starvation and emigration.

The future ability of governments to feed a growing world population will depend to a large extent on our success in understanding the factors controlling the earth's climate and in predicting future changes in it. In this context, the events of the Ice Ages take on a special significance, since a thorough understanding of past climatic changes is essential if there is to be any chance of predicting future trends.

Evidence of our glacial past lies all around us, in patterns of landscape and in the distributions of plants, animals and their fossil remains. Some of the best-preserved remains of Pleistocene faunas have been found in caves, forming an important part of the subject matter of cave palaeontology. Two factors contribute to the accumulation of such fossil remains. Firstly, caves tend to act as natural traps for large animals. Lacking a cave rescue service to haul them out, Pleistocene vertebrates which fell into potholes generally stayed

Fig. 6.1 An exquisitely painted horse from Lascaux Cave in France.

there, their bones adding to the growing pile of earth and rocks which slowly filled up the cave. Some animals used caves as dens for breeding or sleeping. In European caves large quantities of the bones of bats, of Brown Bears (*Ursus arctos*), Cave Bears (*U. spelaeus*) and of Spotted Hyaenas (*Crocuta crocuta*) built up in this way. The latter habitually carried off the bones of their prey to be gnawed at leisure in their den. The chewed remains were left strewn about the cave floor. Secondly, bones in caves have a good chance of being preserved because the cave roof shelters them from weathering processes and lime-laden karst waters do not dissolve them away. Minerals deposits in caves may also preserve the impressions of more delicate, perishable remains, such as the caddis fly wings and maple leaf imprints found in Elderbush Cave, in the Manifold Valley of Staffordshire.

Further evidence of Pleistocene faunas comes from the discoveries of archaeologists. Prehistoric people portrayed their world on the walls of rock shelters and caves. Some of their pictures of extinct creatures such as the Woolly Mammoth (*Mammuthus primigenius*), Aurochs (*Bos primigenius*), Woolly Rhinoceros (*Coelodonta antiquitatis*) and Cave Bear, survive alongside representations of more familiar species, such as the Red Deer (*Cervus elaphus*), Reindeer (*Rangifer tarandus*), Wild Boar (*Sus scrofa*) and Horse (*Equus ferus*). The astonishing accuracy and detail with which extant species were portrayed lends conviction to the archaeologists' reconstructions of the appearance of the extinct forms, based on the record left by those Palaeolithic artists. As a record of former distributions of Pleistocene mammals, cave art broadly confirms what is known in greater detail from Palaeontological evidence, without adding anything of note.

The principal sites of European Palaeolithic art are restricted to an area only about five hundred kilometres square centered on the western Pyrenees, with especially important centres at Les Eyzies in the Dordogne region of France (including the Lascaux cave), the Pyrenees (Niaux and Trois Frères) and along the north coast of Spain (including Altamira). The artists worked in a number of different media. On cave walls and ceilings they made paintings in one or more colours, and engravings. Portable objects such as pebbles, bones, tusks and antlers were also decorated with detailed engravings. Three-dimensional models were fashioned from the clay of cave floors, the finest examples being the Bison group in the Tuc d'Audoubert and the headless clay bear in the cave of Montespan. A few clay models were baked, producing the portable figures of Rhinoceros, Lions and other animals found at Dolni Vestonice in Czechoslovakia.

Few parts of the world can match the British Isles for the variety of Pleistocene fossils found in its caves. The Pleistocene was a time of frequent and profound climatic upheaval, with a constant to-ing and fro-ing of entire biotas as the ice sheets advanced and retreated. In England and Wales, the southerly Hippopotamus (*Hippopotamus amphibius*) and Straight-tusked Elephant (*Palaeoloxodon antiquus*) alternated in late Pleistocene times with such northern forms as the Wolverine (*Gulo gulo*), Woolly Mammoth, Woolly Rhinoceros and Reindeer. Their remains are commonly preserved in caves, sometimes in stratified sequences within a single cave, providing information about the climatic history of the British Isles. However, cave deposits, unless capped by solid travertines, are relatively fragile and transient features, easily eroded away by rejuvenated cave streams during warmer, wetter periods. Consequently, the survival of fossil material in caves becomes progressively more uncommon with age. Deposits of Upper Pleistocene age and younger are common enough in our caves, but earlier deposits are rare.

The earliest Pleistocene cave deposit in the British Isles is Dove Holes in Derbyshire, with a mammalian fauna including the elephant-like Mastodon (*Anancus arvernensis*), Sabre-toothed Cat (*Homotherium latidens*) and Horse. Slightly younger are the Cromerian deposits of the Westbury Cave on Mendip which include remains of Etruscan Rhinoceros (*Dicerorhinus etruscus*), Desman (*Desmana* sp., probably *moschata*), Deninger's Bear (*Ursus deningeri*), Mosbach Wolf (*Canis lupus mosbachensis*), Hyaena and Sabre-toothed Cat. Kent's Cavern in Devon has deposits spanning from the Middle Pleistocene (almost as old as those of Westbury) right through to the present interglacial. The Cave Earth deposit of Kent's Cavern contains a typical Devensian (or Lastglacial period) fauna including Giant Deer (*Megaceros giganteus*, also known misleadingly as the 'Irish Elk'), Woolly Mammoth, Woolly Rhinoceros, Horse, Red Deer, Reindeer, Spotted Hyaena and Cave Lion, all associated with Upper Palaeolithic artifacts. The most important sequence of late Pleistocene cave deposits in the British isles is that of Tornewton Cave in Devon, which was used as a den by Brown Bears during the Wolstonian Glaciation, then by Spotted Hyaenas during the succeeding interglacial – at which time the cave accumulated the remains of Narrow-nosed Rhinoceros (*Dicerorhinus hemitoechus*), Hippopotamus, and Red and Fallow Deer, as well as at least 300 Hyaenas. During the last glaciation, the Reindeer Stratum acquired bones of Woolly Rhinoceros, Horse and Hyaena, associated with Upper Palaeolithic human artifacts. Exactly the same species of large mammals are represented in the Wolstonian

Fig. 6.2 Charterhouse Warren Farm Swallet on Mendip has yielded a wealth of remains from the Neolithic, Bronze Age and Roman periods, including nearly 300 human bones as well as those of Aurochs, Wolf and domestic animals. (Chris Howes)

and Reindeer strata, but the list of small mammals is very different. The lower (older) cold stage is characterized by Steppe Lemming (*Lagurus lagurus*), Snow Vole (*Microtus nivalis*) and two forms of Hamster (*Cricetus cricetus* and probably *Allocricetus bursae*) which are replaced in the younger cold layer by abundant remains of Narrow-skulled Vole (*Microtus gregalis*).

The fact that many of the more spectacular Pleistocene mammals represented in cave deposits are now extinct (probably in a number of cases as a direct result of human activities) makes it difficult in some cases to interpret the climatic significance of their occurrence in a particular site at a particular time. Fortunately the sediments which preserve these enigmatic bones often contain the remains of smaller mammals, birds, and in some cases invertebrates and plants, which are familiar to us today. The distributions and ecological preferences of these species are well known, and make it possible to reconstruct the palaeo-environment in which they must have lived with some confidence.

Mammals with very precise ecological requirements, such as the Beaver (*Castor fiber*), provide more detailed evidence of local conditions. Beaver bones are found in a Lateglacial assemblage in Gough's Cave on Mendip, dated at 12,000 years B.P. and clearly indicate the presence nearby of a large stream or lake, with adjacent deciduous woodland. The same deposit contains remains of the hamster-like Steppe Pika (*Ochotona pusilla*) and Horse, characteristic of open grasslands.

In this brief dip into the vast field of Pleistocene palaeontology, I have been able to mention only a few mammal species found in a small number of the many cave deposits known from the British Isles. For a more detailed account of this fascinating subject, I refer my readers to Antony Sutcliffe's excellent and readable book *On the track of Ice Age mammals* (British Museum (Natural History), 1985).

Unlike the terrestrial mammals, birds during the Pleistocene were less constrained in their geographical movements by the appearance and disappearance of land-bridges and sea-channels, so that their distribution patterns more directly reflect the changing climatic conditions. Fossil remains of species such as the Raven and Eagle Owl which inhabit a wide range of latitudes, and seasonally migrating species, provide little information about palaeo-climates. But some other birds, such as the more sedentary, ground-feeding Grouse and Partridge species, are known to have narrow ecological preferences and so provide precise indications of the conditions where they occur. The Willow Grouse (*Lagopus lagopus*) feeds mainly on the shoots and berries of low shrubs, such as Crowberry and Heather which are typical of moorland or open tundra. The Common Partridge (*Perdrix perdrix*) favours more sheltered conditions and its occurrence alongside Willow Grouse remains in, for example, the 12,000 year-old horizon in Gough's Old Cave on Mendip, indicates a varied vegetation of grassland with some trees as well as moorland. Mallard (*Anas platyrhynchos*) and Tufted Duck (*Aythya fuligula*) bones in the same horizon testify to the many tarns and lakes of this Lateglacial landscape (reinforcing the mammalian evidence of Beaver, Horse and Steppe Pika remains in nearby Gough's Cave). The bones of forest-associated birds found in deposits of Lateglacial/Post-glacial age in caves of the Peak District reveal a succession from tundra to Pine forests (evidenced by remains of Capercaillie, *Tetrao urogallus*) and later to mixed Oak forest (Owls, Doves, Tits, Warblers and so on) which persisted until the area was cleared for agriculture.

Pleistocene survivors and recent colonists

Cave biologists in the British Isles have often found themselves having to defend their interest in what would seem at first sight to be a somewhat second rate cave biota, which lacks the wealth of weird and wonderful forms characteristic of southern Europe and the warmer parts of the USA. The more bizarrely specialized species of those continental caves are often the survivors of very ancient lineages and have been termed "palaeotroglobites" by Vandel. Jeff Jefferson, in *The Science of Speleology* (1976), wrote that

> "it is doubtful whether the British cave fauna contains a single terrestrial palaeotroglobite ... many of our terrestrial cave animals ... have not been sufficiently isolated to have diverged to the point at which their specific distinctness is beyond question."

At first sight this assertion seems curious, as one might presume that the British Isles have been furnished with habitable caves for much the same length of time as anywhere else in the world. In fact this is not the case at all, for during the Pleistocene epoch most cave areas in Britain and Ireland became ice-covered at some time or another, cutting off the food supplies of the species which inhabitated them, causing them to die out.

Many of our terrestrial cavernicoles belong to groups more typical of the taiga-, or tundra-covered terrains of northern Europe than of the temperate woodlands which form the climax vegetation of the zones they now inhabit. Such species are 'cryophilic relicts' – survivors of a late Pleistocene periglacial fauna which occupied a band to the south of the ice sheets which then covered much of northern and central Britain and Ireland. As the glaciers retreated, and the vegetation bands moved northwards in their wake, so this fauna gradually followed, finding refuge on mountains and in caves as the climate warmed and new, warm-adapted communities invaded from lands further south. Since the last glacial retreat was, on a geological timescale, a mere blink ago, any such forms would presumably not yet have had time to develop the distinctive morphological features associated with longstanding cave occupation, and would not show the degree of local speciation to be found among the more ancient cave relicts of other parts of the world. The collembolan *Onychiurus schoetti* seems to be an example of such a recent cavernicole, being recorded from soils and leaf-litter in Norway, but only from cavernous habitats in Britain and Western Europe. The genus *Porrhomma* which includes our only troglobitic spider, is a markedly northern group, characteristic of mountains and sub-arctic vegetation. The cavernicolous fungus-gnat *Speolepta leptogaster* and carabid beetle *Trechus micros* also appear to occupy caves in Britain close to the southern limit of their range.

Not all terrestrial cavernicoles in our islands are cryophilic relicts. The troglobitic springtail, *Pararrhopalites patritzii*, found in caves in Devon and nowhere else in the British Isles, is known to have a southern distribution on the European mainland. It would appear that it is a 'thermophilic relict' which colonized southern Britain during a warm Pleistocene interglacial period (or possibly earlier) and was driven underground by the deteriorating climate associated with a period of glacial advance. It is now confined to subterranean habitats at the northern limits of its distribution.

Active colonization or recolonization of our caves still appears to be in full swing. I have earlier mentioned the spectacular above-ground invasion by the snail *Hydrobia jenkinsi*, first of estuaries and then of freshwater streams and its seemingly recent entry into caves. Another intriguing instance of a similar phenomenon is seen in the tiny blind isopod *Trichoniscoides saeroeensis*. It was first found on a sea shore near Gothenburg in southern Sweden in 1922, in deep humus under cast-up weed at the high water mark and has since been

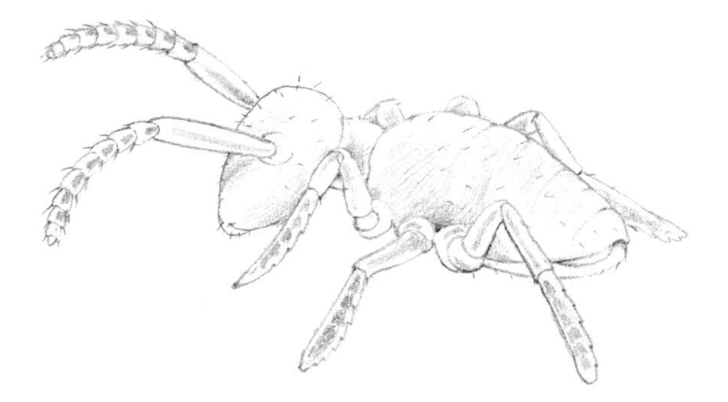

Fig. 6.3 The springtail *Pararrhopalites patritzii* – a refugee from warmer climes.

recorded in similar situations along the coasts of Denmark, Northern France, the islands of Gotland, Oland and Bornholm, from a seaside mine in Lancashire, and from the Poulnagollum-Poulelva and Doolin Cave systems in Co. Clare. Vandel concludes from the available evidence that this is a warm-loving, expansive species which has spread only recently in Quarternary times from an original dispersal centre in south-west France. There was probably no direct connection between the western part of the English Channel and the Danish and Swedish coasts until about 8000 years ago. By then, the climate had recovered from the last glacial phase, and *T. saeroeensis* would have been able to migrate northwards and north-eastwards along the French coast and then westwards along the south coast of England. When the isthmus which had joined Sussex and Kent with North-eastern France became finally submerged, the way was clear for migration of many species to Denmark and Sweden. However, the spread of *T. saeroeensis* into Ireland is not so easy to explain. As far as is known, Ireland has not been directly joined to Britain at any time when small creatures needing moderate warmth could have flourished. It would seem most unlikely that the isopod could have come in during the last interglacial warm period over a million years ago (when sea levels may have been about 200 m above their present level) and then have survived through several severe glacial periods. However, as a littoral inhabitant, it may have somehow drifted across to the Irish coast from France in more recent times and become established.

It is at first sight puzzling that all the British and Irish records of *T. saeroeensis* should be from caves, whereas those from the continent are mainly from beneath seaweed. It may be that in this species we are witnessing the invasion of a novel habitat by an expansive, adaptable species, in much the same manner as proposed by Frank Howarth for the cave-invading Hawaiian crickets *Caconemobius* spp. – whose littoral ancestors drifted across the ocean through the islands of the Hawaiian chain, became established on the sea coasts where they landed and migrated up into caves as a predictable expansion of their range. As with the Hawaiian cave crickets, *T. saeroeensis*' move into caves has been accompanied by a considerable extension in its altitudinal range, from sea level to over 200 m elevation in Co. Clare. Evans and Evans have suggested that the little isopod may have been unwittingly introduced by farmers to the high ground near the Poulnagollum-Poulelva cave system, in among the seaweed which is traditionally collected along the shores of Galway Bay to spread on the land as a fertilizer. As *T. saeroeensis* is probably adapted principally to living among the interstices of sea-shore cobbles, it is easy to visualize how it might find its way down into the mesocavernous spaces and cave passages of the High Burren.

In its apparent ecological opportunism, *T. saeroeensis* illustrates a feature of cavernicolous faunas which is too often overlooked. Despite their morphological peculiarities of eyelessness and depigmentation, these animals are by no means necessarily on the retreat, in ecological terms, but may on the contrary be very successful, aggressive and expansive species, although of course they cannot live above ground. Cavernicoles survive in the rigorous world of caves precisely because they are extremely fit, or well-adapted, to cope with the conditions found in such places.

There is clear evidence that in most countries within temperate latitudes, many of the most specialized cavernicoles are relicts of an ancient tropical fauna which probably invaded subterranean niches during the first half of the Tertiary. Some aquatic troglobites probably evolved from ancestors which entered caves directly from surface habitats, perhaps washed in by flooding – a process which may be witnessed in its earliest stages today in the case of *Gammarus pulex*. For such forms to have developed troglobitic features, they would have had to be genetically isolated from any surface populations for a long period, as would happen with a drying of the climate to the point at which surface streams ceased to flow for a substantial part of each year. Others seem to have entered freshwater caves from the equivalent marine habitat, via brackish coastal groundwaters (the 'anchialine' habitat). There is general agreement among the specialists that most subterranean amphipods have colonized fresh groundwaters in this way. In the case of our own *Niphargus* species and *Niphargellus glenniei*, the distributional evidence points to a centre of origin in a southerly part of central Europe. According to Ruffo, the orig-

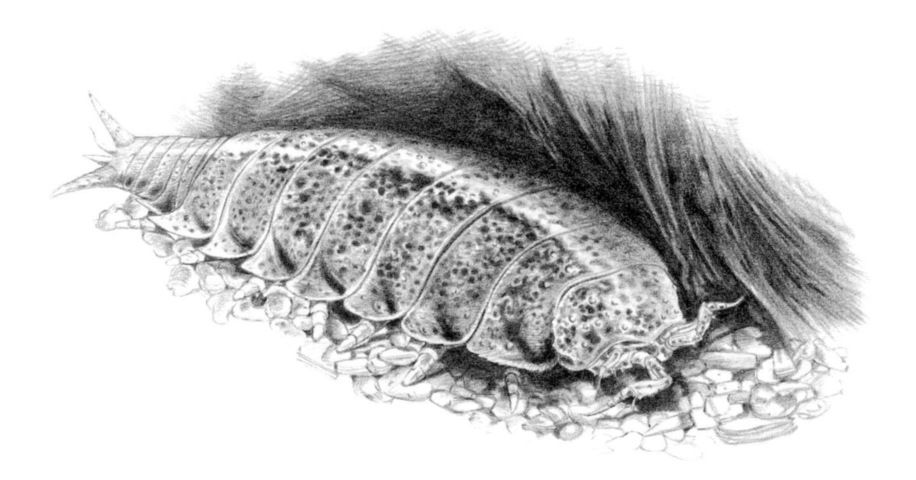

Fig. 6.4 Trichoniscoides saeroeensis – in the process of expanding its range into caves in Britain and Ireland.

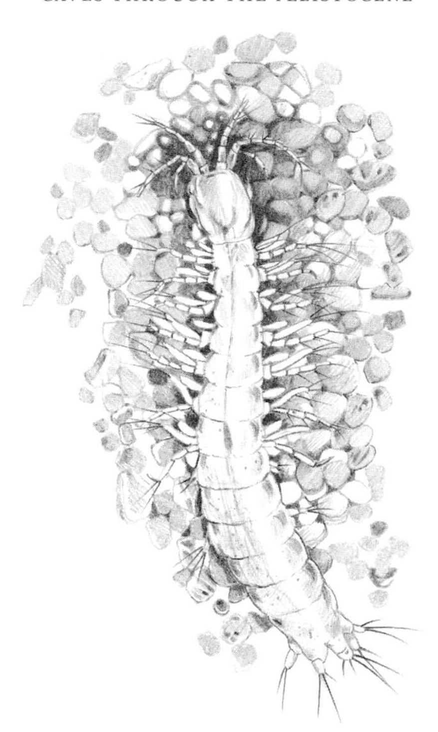

Fig. 6.5 Bathynella stammeri, a 'living fossil' found in groundwaters throughout Britain.

inal saltwater-to-freshwater colonization may have occurred in Miocene times around the perimeter of a sea which became progressively more brackish. Once established in brackish groundwaters, various lines of *Niphargus* radiated into sweet groundwaters and then spread across Europe to their present limits. The *Crangonyx* group of amphipods is much more widely distributed – our own *C. subterraneus* being found throughout Eurasia, South America and South Africa, suggesting that these amphipods have been in fresh water for an even longer time – perhaps since the Trias, before the separation of the ancient southern continent of Gondwana from Laurasia.

Among the isopods, the family Asellidae has probably been a freshwater group since the Mesozoic, according to Birstein, and the genus *Proasellus* is thought to have enjoyed an even wider geographical range during the Tertiary, but has since been squeezed out of many former strongholds by more recently evolved competitors. *Bathynella* belongs to a very ancient group, the Syncarida, which were originally described from fossil forms. The syncarids which are found in rocks of the Lower Carboniferous era are all marine forms, but during the Upper Carboniferous and Permian eras, they invaded freshwaters and became widespread in Europe and the Americas (then part of a single supercontinent, Pangaea). *Bathynella* retains many primitive features of the early Syncarida and is in this sense a 'living fossil', and one of the very oldest members of the British fauna.

During the latter part of the Pliocene and early Pleistocene, most of the lower-lying parts of Europe, including our islands, were under the sea. The Pleistocene saw fluctuating sea levels which resulted in the British Isles being connected, cut off and re-connected several times with continental Europe. Our present isolation dates back a mere 8500 years, and the wave-cut platforms of Pembrokeshire and elsewhere, clearly show that there has been a considerable overall lowering of sea-level since the early Pleistocene. The effects of the high sea stance of the late Pliocene would have been to break up the familiar landscape of modern Britain into a number of islands. In the south, the karstic regions of Devon, parts of Ireland, Wales, Mendip and the chalk downlands of south and south-east England would all have been separate islands. Opportunities for dispersal between islands would have been particularly limited in the case of subterranean species and these would have tended to differentiate during the period of separation and could be expected to retain a measure of distinctness even after the islands had rejoined. This could explain the marked differences between the subterranean faunas of Devon and Mendip, which both lie south of the maximum glacial limit, and so might otherwise be expected to have retained rather similar cave faunas during the Pleistocene.

It is noticeable that in Europe, the distributions of ancient aquatic troglobites extend considerably further north than do those of terrestrial forms, suggesting that the former were better able to withstand the climatic vagaries of the Pleistocene. Evidence to support this contention comes from Holsinger's studies of eleven species of subterranean amphipods that occur north of the southern limit of Pleistocene glaciation in eastern North America. Most, if not all of these are primarily associated with interstitial groundwaters, rather than specifically with caves, and Holsinger provides convincing evidence that at least four of these survived glacial periods in deep groundwater beneath the ice, and that the chances of survival under the ice increase with increasing depth of the habitat. As terrestrial cavernicoles are necessarily confined to shallower depths (above the water table), it seems reasonable to suppose that they should have been more severely affected by glacial and periglacial conditions, and this probably accounts for the relative scarcity of terrestrial troglobites in Britain and Ireland.

The northerly limit of distribution of several of the *hypogean* Crustacea found in Britain and Ireland corresponds fairly well with the southernmost limits of glacial advance during the Pleistocene, although *Niphargus aquilex* and *N. kochianus*, which inhabit the more superficial groundwaters stray rather further north of this line. A similar distribution pattern is seen in some European hypogean amphipods and in many other groups of troglobites. Ruffo has argued that this pattern results from the annihilation of the subterranean fauna in areas affected directly by ice during the Pleistocene and the subsequent recolonization of those areas via diffuse subterranean routes. The more rapid colonizers are those species which live in more superficial groundwaters.

While this is an excellent model as far as it goes, it fails to explain how some Crustacea characteristic of deep groundwaters should occur actually further north of the glacial limit than their shallow-groundwater cousins. An example is the occurrence of *Niphargus fontanus* in South Wales in areas which have been repeatedly glaciated, including during the latest 'Devensian' glacial advance. Moreover, the model does not explain why the distributions of troglobitic aquatic Crustacea as a group extend further north in Europe than do those of most terrestrial troglobites. These observations suggests to me that aquatic troglobites may actually have been able to survive to some extent be-

Fig. 6.6 The maximum glacial limit in the British Isles and distributions of two cave Crustacea, *Niphargellus glenneii* (●) and *Niphargus fontanus* (○). *(After G.T. Jefferson)*

neath the ice during Pleistocene glacial periods. Support for this view is provided by Jefferson, who argued that the substantial differences between the populations of *Proasellus cavaticus* on Mendip and in South Wales points to a long separation between the two populations. This situation could only have been maintained throughout the Pleistocene if the Welsh *P. cavaticus* had survived beneath the ice. Further evidence is provided by Holsinger's studies of eleven species of subterranean amphipods that occur north of the southern limit of Pleistocene glaciations in eastern North America. He demonstrates that at least four of these must have survived glacial periods in deep groundwater beneath the ice, and suggests that the chances of survival under the ice increase with increasing depth of the habitat. Because terrestrial cavernicoles are necessarily confined to shallower depths (above the water table), Holsinger's observations predict that they should be more severely affected by glacial and periglacial conditions. This would to a large extent account for the relative scarcity of terrestrial troglobites in Britain and Ireland.

If aquatic troglobites were able to survive below the ice cover, it seems that this was only possible in areas like East Anglia, South Wales and possibly parts of Ireland, not too far from the edge of the ice sheet, where some organic material may from time to time have found its way into groundwaters in sufficient quantities to sustain life. Rates of post-glacial recolonization have varied greatly among different species of subterranean Crustacea. *Niphargus aquilex* has reached Humberside and and the Tees. The sprightly *Bathynella*, shrugging off its enormous age, has made it all the way to Scotland. But other species have been slower off the mark – *Crangonyx subterraneus* and *Niphargus fontanus* have barely advanced beyond the limit of their distribution at the close of the Pleistocene.

Troglodytes

In caves, our Palaeolithic ancestors found a welcome shelter from wind and rain and from wolves and other wild beasts. Free-standing houses are a surprisingly recent invention, and even today, we need look no further than France to find an estimated 25,000 people still living in caves and rock dwellings, while in Britain, as David Kempe records in his book *Living Underground* (Herbert Press, 1988), the cave-houses around Kinver Edge, north of Kidderminster, were lived in until the 1970s – the last being vacated in 1974.

The term 'troglodyte', as applied to modern cave-dwellers, has its origin in the *Trogodytae*, descendents of the Cushites who lived in about 2250 BP along both sides of the Red Sea. The Roman Pomponius Mela is quoted in Rees' *Encyclopaedia* of 1817 on the customs of the original troglodytes. It appears that they ate serpents and lizards and dwelt in underground caves. They did not properly speak, but hissed and shrieked like owls or bats. They were swift of foot, kept women in common without marriage and were governed by tyrants. They buried their dead by binding their necks, knees and heels together, pelting them with stones and, amidst communal laughter, placing a goat's horn upon the corpse. Old men, when not killed by the tribe, frequently committed suicide by tying a bull's tail around their neck and being dragged to death. Anyone reaching the age of sixty was killed out of hand. In short, they were as charming a bunch of people as one might hope to meet.

Recent evidence from Africa suggests that the fire-using *Homo erectus* may have originated as long ago as 1.8 million years BP. The earliest evidence of cave habitation by the species (dated around 700,000 BP) occurs, according to

Fig. 6.7 Home of an English troglodite – the rock house at Kinver.

Fig. 6.8 A neolithic human jawbone lies among boulders in the Bone Chamber of Charterhouse Warren Farm Swallet. (Chris Howes)

Kempe, in the Acheulian le Vallonet cave on the French side of the Italian border. In the Zhoukoudian caves near Beijing in China, chert chipping tools typical of *Homo erectus* have been found in red clays dated at about 500,000 BP. The nearby Zhoukoudian Great Cave is famous as the site which yielded the first fossil skull of 'Peking Man', originally named *Sinanthropus*, but now attributed to *H. erectus*.

Homo sapiens, of an archaic type, seems to have begun to leave traces in Europe around 400,000 BP. The first widespread and successful subspecies was the heavy-browed *H. sapiens neanderthalensis*, who was ensconced in various cave sites around 150,000 BP, close to the start of the Middle Palaeolithic period. The middens, or refuse heaps of Neanderthal People suggest that, in common with many present day hunter-gatherers, they subsisted on a mainly vegetarian diet, supplemented by small mammals, reptiles, crustaceans and occasional larger prey. They may have kept meat in storage pits covered with stone slabs to keep off scavengers and cooked it on hot stones. Their tools were crudely made and progressed little in 100,000 years. Some of the earliest Neanderthal cultural remains in the British Isles are a number of hand axes and African-style cleavers excavated from a sea cave at La Cotte de St Brelade in Jersey, which at the time of deposition, about 150,000 BP, would have been joined to the European mainland. Neanderthals survived in Britain up until about 30,000 BP and may have been finally replaced by members of a different hominid subspecies, destined to become the most succesful and destructive creature the earth has ever known.

The earliest fossils attributed to *H. sapiens sapiens* have been dated by associated flints, bones, or sediments to 100,000 BP in South Africa (Klasies Caves and Border Cave) and Israel (Qafzeh and Skhul Caves). The European early modern 'Cro-Magnon Man' was better-armed and altogether more innovative than the Neanderthals he rapidly began to oust following his arrival in Europe sometime around 40,000 BP. His 30,000 year rule saw a cultural explosion in beautifully-fashioned tools made of bone, antler, wood and stone, in representational cave paintings, etchings, carvings and even decorative jewellery.

The evidence for the presence of Cro-Magnon Man in the British Isles is far richer than for his Neanderthal predecessor. The first traces of his presence here come from a series of leaf-point flints and a jaw fragment found at Kent's Cavern, Torquay, and dated respectively at 35,000 and 31,000 BP. The famous 'Red Lady' burial at Paviland on the Gower (actually the skeleton of a 25 year-old male) has been dated to 26,000 BP. Paviland is perhaps the richest early Upper Palaeolithic site in Britain and Kempe's description of the Cro-Magnon habitation there makes it sound positively luxurious. Apparently the cave was dry, well-lit, with a natural chimney to carry off the smoke from the fire and a rocky platform in front equipped with natural rock seats. The large midden on its red loam floor contained the bones of Horse, Bison, Woolly Rhinoceros, Hyaena, Woolly Mammoth and Cave Bear. The famous cave burial contained flint implements, sea-shells pierced for wearing, small ivory rods and an ivory ring, in addition to the ochre-smeared body itself.

Occupation of the Paviland site seems to have come to an end with the deterioration of the climate at some time around, or soon after, 25,000 BP, and from then until about 13,000 BP much of Britain appears to have been largely uninhabited. This interval corresponds roughly to the coldest part of the Last-glacial, or Devensian period. At the height of the glacial advance, the southern limit of the ice sheets ran across Ireland from the Shannon to Wicklow, along the South Wales coast, missing out Pembrokeshire, the Gower and South Glamorgan, up along the Welsh-English border and on up to Manchester and from there across to the North Yorkshire coast, with a southward dip around the Pennines.

As the glaciers retreated around 13,000 BP and the climate gradually warmed, people began to move back into southern Britain, following the herds of large herbivores which provided them with the meat, skins and other products on which hunter-gatherer societies still depend today. Among the new sites to be colonized were Creswell in Derbyshire, Nanna's Cave on the Gower, King Arthur's Cave on the Wye, Cefn in North Wales and Kirkhead in Cumbria, and later, even more northerly Mesolithic sites at Star Carr in South Yorkshire, Victoria Cave in North Yorkshire, MacArthur Cave at Oban in Scotland and the valley of the River Bann in the far north of Ireland.

During the Mesolithic, and increasingly through the Neolithic and Bronze Ages, caves gradually fell out of favour as habitations, but continued to be used extensively as places of burial. A typical example is Dowel Cave in the Peak District of Derbyshire, where a narrow passage seven metres long was sealed by portal slabs and two internal blockings, and contained the buried remains of ten individuals. The sepulchral use of natural cave passages and shafts is also found in the north-west in Yorkshire and Lancashire, but not on Mendip.

7

The Future of Caves

Cave conservation and the caver

"The cave ecosystem is perhaps the only truly natural ecosystem available for study in a country such as Britain."

So wrote Jeff Jefferson, in a report to the Nature Conservancy Council in 1979. David Judson, the British Cave Research Association's Conservation Officer, and Environmental Consultant Dr Laurie Richards sounded a more cautionary note in their 1989 review of the state of cave conservation in Britain:

"Caves provide a unique environment and a unique challenge to the conservationist ... no other features on the earth's surface are so vulnerable and at the same time are subjected to such concentrated human pressure."

These are striking statements. If justified, they must surely mean that the conservation of cave communities should be given a much higher priority than hitherto.

Judged by the standards of surface habitats, caves are nationally rare features. They are irreplaceable on an ecological time scale. Caves which have reached the stage where they are being infilled by speleothems and sediments are often extremely fragile. Some elements of our cave-associated fauna, such as Horseshoe Bats, are known to be vulnerable to human disturbance, and have declined alarmingly in recent years. Other, more cave-dependent species seem to be present within caves at extremely low population densities; they suffer from the double vulnerability of being rare within a habitat which is itself rare.

In broad terms, there are two types of threat to cave environments: those which come from the use of caves by people, including physical erosion, pollution and vandalism; and those which stem from human activities outside caves, including quarrying, construction works, waste disposal and various agricultural operations. It used to be widely held among cavers that external activities presented the greatest threat to caves, but the concensus is now that cavers themselves are often the principal agents of destruction. Willie Stanton, a veteran Mendip cave explorer, writing in 1982, summed up the problem as follows:

"The present position is one of steady deterioration in all caves ... Of Mendip's 550 caves, 31 had what I would call fine grottoes when they were discovered, beginning with Lamb Leer in 1674 and ending with Manor Farm in 1973. Now 17 of those 31 no longer contain significant stalactite scenery ... The destruction in Swildon's, G.B., Stoke Lane, Lamb Leer and smaller caves such as Sidcot, Loxton and Rickford Farm has been colossal ... And there lies the rub: if you don't realise what was there, you don't miss it." (*Cave Science* 9:176–7).

A total of 48 cave sites, representing the best underground scenery in mainland Britain, have been, or are presently being designated as 'Sites of Special Scientific Interest' (SSSIs) by the Nature Conservancy Council under the terms of the Wildlife and Countryside Act 1981 (the NCC itself has now been disbanded and replaced by three regional bodies: English Nature, Scottish Natural Heritage and the Countryside Council for Wales). However, SSSI status, although providing some safeguards against external pressures on caves (for example by requiring landowners to consult with the regional bodies prior to undertaking 'Potentially Damaging Operations', such as ploughing within the hydrological catchments of protected caves), has not proved effective in influencing the behaviour or attitudes of cavers towards conservation of the caves themselves. Most designated caves are visited exclusively by cavers, who comprise the only group of people with the capability to explore, study, monitor or police the underground environment. It would therefore seem sensible that they – the cavers and the organizations which represent them, the National Caving Association (NCA) and British Cave Research Association (BCRA) in Britain, and the Speleological Union of Ireland (SUI) – should be encouraged to assume the major responsibility for cave conservation.

The close relationship between the level of usage of a cave and the amount of damage to features within it has long been recognized. A report by Wilmut in 1972 (*Caves and Conservation*, NCA, Birmingham) showed that some features within 12.7% of all known caves, and 26.5% of major caves in Britain had already been damaged or destroyed by that time. Since 1972, caving has continued to increase in popularity and, as Judson and Richards point out, "it is inevitable that further damage through increased usage will have occurred".

Following the 1972 report, the NCA have taken a number of initiatives aimed at promoting cave conservation, including the publication of a *National Caving Code*, and production of an educational film, *Lost Caves*. More recently,

Fig. 7.1 Stalagmites broken by vandals, Ogof Craig a Ffynnon. (Chris Howes)

the NCA, BCRA and NCC set up 'Cave Liaison Groups' in each major caving area in an attempt to improve communications and cooperation between themselves and with local landowners. The SUI has followed a similar path in appointing a 'Conservation and Access Officer' and three regional 'wardens' based in Clare, Fermanagh and Tipperary. However, there is no clear evidence so far to suggest that these initiatives have been successful, or of any halt in the rising tide of destruction within caves.

External threats to caves, groundwater pollution and public safety

A report prepared for the NCC by Paul Hardwick and John Gunn of the Limestone Research Group at the former Manchester Polytechnic discusses the various human operations in karst areas in Britain which are known to have affected caves. Top of the list comes rubbish tipping (35 sites affected), followed by farm tipping (16 sites), infilling of caves (14 sites), open ditching (seven sites), spillage of oil, pesticide or industrial effluent (six sites), sewage ingress (five sites) and slurry and silage disposal (four sites), with other operations affecting a further six cave sites.

Holes in the ground have always been used to dispose of unwanted and unsightly rubbish. Karst areas abound with unofficial natural landfill sites in the form of dolines and open cave entrances, as a walk across any of our limestone areas will confirm. The use of caves as charnel houses for the disposal of livestock carcasses seems particularly prevalent in the Burren of Co. Clare, and must pose a considerable health risk. Groundwater pollution resulting from general dumping of domestic and animal wastes is a problem in many countries.

In Britain, 148 licenced landfill sites are situated within areas of Carboniferous Limestone. Of these, no less than 67 are allowed to take industrial wastes. Edwards and Smart, in a review of such sites, suggest that this is due to the remote location of most of the sites, since it is only in the case of the more dangerous toxic wastes that the high transport costs of such materials is justified. The same authors reported five instances where unidentified leachate from such sites is known to have contaminated underground watercourses, in one case travelling over two kilometres and causing gross pollution of a rising downstream. In addition, leachate from landfill sites has turned up in boreholes at six further sites in England and Wales, suggesting widespread contamination of the respective aquifers. The evidence suggests that industrial wastes are likely to have played a major part in all 11 of the above leachate incidents. Given the nature of karst drainage, it seems quite unbelievable that permission should have been given for the dumping of harmful industrial wastes at nine licenced sites on Mendip alone – an area which is supposed to have one of the strictest aquifer protection policies in Britain.

A study in Ireland of disposal of silage and farmyard slurries resulting from intensive livestock rearing found that up to 100,000 m^3 of wastes per year were discharged from each feedlot, often directly into sinkholes, soakways and old quarries, often in fissured limestone aquifers. In addition to this deliberate pollution, there are documented instances of accidental pollution of karst groundwaters in Ireland caused by poor storage of farm slurries and silage.

Studies made in Germany have shown that the insecticide lindane, used to control Bark Beetle in coniferous forest, is retained very efficiently by karst soils, although a small proportion (0.3–2.5 %) may reach groundwaters very rapidly after rain. Nitrate fertilizers behave quite differently, being concentrated and retained within the karst aquifer, particularly when the recharge is by autogenic percolation. In a recent study, Hardwick found that percolation water dripping into cave passages in P8 cave in Derbyshire contained ten times more dissolved nitrates than drip water entering Yorkshire's White Scar Cave (means of 40mg/l and 4mg/l, respectively). The former cave underlies agricultural land which is treated with sewage sludge as a fertilizer, while the latter is overlain by bare limestone pavements. It appears that suspended particles of silt soak up ammoniacal nitrogen and phosphorus, and thus act as a nutrient store. The accumulation of fertilizers in groundwater may be responsible for turning the once crystal-clear waters of many of the *cenote* lakes of southern Australia into an undrinkable algal soup.

In Britain, the discharge of agricultural wastes, noxious or toxic chemicals into watercourses is a criminal offence, but the nature of karst aquifers often makes it difficult to pinpoint the source of contamination. One of the few successful prosecutions followed the pollution by silage liquor of Ashwick Grove Springs on Mendip in 1969. The liquor was discharged into a land drain feeding a stream sink, causing a forty-fold rise in phenol concentrations in the local drinking supply. In another Mendip example, creamery waste was illegally dumped in a quarry over a kilometre from Holwell Rising, causing de-oxygenation up to 1.5 km downstream. Perhaps the worst pollution event on Mendip occurred in 1967, when 227,000 litres of waste oil contaminated with cyanide and toxic metals were illegally dumped by a waste disposal firm into Nedge Hill Hole. In 1971, the cave was still too polluted to enter and the entrance was infilled by the landowner. In Devon, access to Coombesend Cavern has been rendered hazardous by the dumping of 'obnoxious wastes' in the quarry in which one entrance to the cave is situated, and another entrance has been blocked since 'obnoxious gases' are said to be leaking into the system. In the Peak District, oil pollution from an unknown source is known to have caused temporary illness to a number of cavers visiting Peak Cavern. Vapour from the oil caused choking, nausea and vomiting. Similar problems have been experienced in nearby P8 cave, in this case attributed to elevated CO_2 levels produced by oxidizing sewage sludge in a confined cave passage.

The pollution of karst groundwaters by untreated domestic sewage constitutes a serious health hazard in parts of the world where treatment facilities are lacking and where pathogenic contamination of public water supplies by sewage may lead to serious disease epidemics. For example, between 1925 and 1973, 75 typhoid-paratyphoid epidemics affected populations in the karst regions of Slovenia. Moreover, in a report published within the last decade, Larson and Larson stated that most water in caves in urbanized karst areas of the United States should be considered suspect and a potential transmitter of diarrhoea, hepatitis, typhoid and other diseases. It may seem surprising that such a situation can occur in a rich, developed nation even today, yet the reality is that many people living in karst areas still have little concept of the nature of the underground drainage which supplies their drinking water and which conveniently removes their waste. An interesting example, cited by Waltham in his 1974 book *Caves* (Macmillan), concerns the Ballymacelligott Caves near Tralee, Co. Kerry. The caves are fragmented by collapse into a number of short sections spread out over 1.5 km along a valley, with seven known entrances, excluding the top sink and the bottom rising. It was only realized that all nine openings were hydrologically connected following a visit by cavers, who explored much of the system for the first time. At the third entrance from the top, cattle used the stream for drinking and other purposes. At the sixth opening down, the stream was used for a house supply, and gathered the lavatory sewer which fed directly into it. The seventh opening was again used by cattle. At the bottom rising the stream flowed out of fissures in the limestone and was used to supply a farm, a pair of cottages and a creamery. This is a level of water recycling which the Thames Water Authority might envy, except that no natural filtration or treatment of any sort had intervened along the chain. One suspects that any residents who kept their health owe a substantial debt to the cavernicolous community which must have biologically cleaned the water to some extent during its underground journey.

Pollution by human sewage has been reported in Mendip caves on a number of occasions, and has been blamed variously on poorly maintained septic tanks and broken sewer pipes. Apart from the direct hazard to health, sewage in caves is frequently associated with disease-carrying animals such as rats, whose role in the transmission of the pathogen responsible for Weil's Disease was discussed in Chapter 4.

Almost 40% of the 1.2 million tonnes of sludge dry solids produced in Britain each year from the treatment of sewage is applied to agricultural land as a cheap fertilizer. Concern over the pollution of limestone groundwaters recently led one Regional Water Authority to suspend supplies of sludge to farmers in the Derbyshire Peak District. Unfortunately the ban was not effective, as Hardwick and Gunn found when they subsequently detected groundwater pollution in P8 (Jackpot) cave, near Castleton, resulting from application of sludge to a field overlying the cave. The farmer had simply (and quite legally) found himself another supplier.

In Kentucky, a recent study has blamed bad ploughing techniques for soil erosion into the Lost River karst drainage, resulting in the underground deposition of several metres of sediment, with its associated pesticides, nutrients, organic and inorganic detritus. The sediment causes an increase in water treatment costs, destroys the habitat of 'desirable aquatic organisms', impairs the dissolved oxygen balance, raises the background level of nutrients and chemical pollutants and may alter the carrying capacity of the aquifer, resulting in problems with drainage and an increased risk of flooding.

Flooding in caves is an occupational hazard faced by cavers, who generally take note of weather conditions before venturing underground and behave appropriately. However, natural flooding can be exacerbated by human activity, with potentially dangerous results. Oliver Lloyd reported a 'near-miss' incident in 1978 in Stoke Lane Slocker on Mendip, when automatic activation of pumps in an adjacent limestone quarry during a rainstorm added to the already swollen cave stream, causing a massive flood pulse which trapped eight cavers. They could not be rescued until the pumps were switched off.

Impact of human activity on cave faunas

In their recent report on the effects of agricultural operations on caves in Britain, Hardwick and Gunn (1990) note that "impacts of water quality changes on cave ecological systems have rarely been considered". As far as I know, there has not been a single such study in any cave area in the British Isles. This is surprising in view of the obvious role played by subterranean communities in restoring and maintaining the quality of groundwater drinking supplies. Fortunately, a few such studies have been conducted in other countries (notably the USA) and the results of these serve to illustrate some of the probable biological impacts of similar operations conducted in our islands.

Toxological assays are often used to assess the effects of pollutants on stream communities. A standard technique involves determining the LD_{50} concentration of a pollutant at which 50% of the tested sample of organisms will die in a specified time. This value often depends on 'Application Factors' (AF), such as temperature or pH of the water, which vary the toxic effect. In nature, all kinds of AF factors come into play, including the presence of naturally occurring chemicals, or of other pollutants (the so-called 'chemical cocktail' effect). Laboratory studies may frequently disregard important AFs, and experimentally-determined 'safe levels' of toxic wastes should therefore always be viewed with caution. In the USA, toxicological studies by Bosnak and Morgan and Barr have found, on the one hand, that non-cave dwelling species of *Asellus* isopods were more sensitive to cadmium and zinc than their cave relatives, and on the other, that the hypogean isopod *Asellus stygius* was more sensitive to nickel, cadmium and chromium, than the epigean isopod *Lirceus fontinalis*. In another study, Bosnak and Morgan found that the hypogean isopod *Caecidotea bicrenata* was four times more tolerant to cadmium than another hypogean isopod, *C. stygia*, probably because the former is exposed to high cadmium levels in its natural habitat. This underlines a general point, that organisms tend to be tolerant of substances to which they are routinely exposed, but less tolerant of unfamiliar chemicals.

To date, only a single troglobitic species has been formally recognized as being in danger of biological extinction as a result of groundwater pollution. The rare and local Kentucky Cave Shrimp, *Palaemonias ganteri,* is listed as an Endangered Species by the US Government, which has made a formal commitment to construct a regional sewage treatment plant to protect the species' habitat and associated fauna from the organic pollution which is its major threat.

Two field studies in the USA in the last decade have documented major cave ecosystem disasters from chemical pollution. In 1979, Brucker found over 100 dead or dying cave crayfish in the Hawkins River of Proctor Cave, Kentucky. The river smelled strongly of gasoline which was presumed to come from a storage tank at an overlying petrol station. An even more dramatic event was documented by Vandike and Crunkilton in Missouri, where 80,000 litres of liquid ammonium nitrate and urea fertilizer leaked into the underground catchment of the Meramec Spring from an agricultural pipeline. Dissolved oxygen in the phreas converted the dissociated ammonium ions to nitrite and then to nitrate, leading to a catastrophic deoxygenation of the spring water. About 1000 Southern Cavefish (*Typhlichthys subterraneus*) and 10,000 Salem Cave Crayfish (*Cambarus hubrichti*) are estimated to have perished. An interesting conclusion to be drawn from this disaster is that chemical substances not normally considered to be toxic to aquatic life can, nevertheless, create major problems within cave ecosystems.

Studies of the impact of increased sedimentation on cave stream ecosystems have been conducted in the USA and New Zealand. In Mammoth Cave, Kentucky, silting caused by human activity led to the loss of pool and riffle sequences essential to a specialized isopod, *Caecidotea* sp.1, and to a general decrease in all isopod numbers, depriving other species, such as the troglobitic crayfish *Orconectes pellucidus*, of one of their principal food sources. The New

Fig. 7.2 Rubbish tipped into a doline on Llangynidr Mountain. (Chris Howes)

Zealand study is of particular interest, as it was funded by the state run Tourist Hotel Corporation which operates the Waitomo Show Caves. The objective was to safeguard a species of cavernicolous fungus gnat, *Arachnocampa luminosa*, whose 'Glow-worm' larvae constitute the principal tourist attraction of the Waitomo Caves. The siltation study ran in parallel with an ecological study of *Arachnocampa* and the results illustrate the complex interactions even within a 'simple' cave ecosystem. Silting of the Waitomo Caves was caused by increased soil erosion resulting from road and trail constructions and from deforestation in the catchment of the underground river, compounded by artificial raising of water levels at the resurgence to improve access by tourists in boats.

Changes in the cave river resulted in a decrease in populations of chironomid midge larvae, whose adult stages emerge in the cave, where they form the main food supply of the predatory Glow-worm larvae. At the same time, changes in the air-flow and microclimate of the caves, caused by artificial blocking of cave entrances, increased the susceptibility of the Glow-worms to fungal pathogens. Having identified the problems, a rehabilitation and management programme was implemented which led to a recovery of the Glow-worm population, and a consequent reprieve for the local tourism industry.

Studies of sedimentation effects on cave faunas are relevant in the British Isles, where ploughing of soils overlying karst landscapes is undoubtedly contributing to soil erosion and to sedimentation of underground habitats and particularly of mesocavernous cracks. Howarth has discussed the implications for Hawaiian cavernicoles of the infilling of the mesocavernous habitat of lava flows by soils. He considers this to be the principal natural cause of extinction of cavernicoles in the older Hawaiian islands. Although confirmatory evidence is lacking, it is likely that this process has already caused, and is continuing to cause, substantial damage to our own cavernicolous fauna, which, as we have seen in earlier chapters, is likely to be primarily adapted to, and dependant on, such habitats.

Another threat to cavernicoles comes from the use of caves as water reservoirs. Damming of karst springs is frequent in China and parts of Central Europe. Generally, no account is taken of the fate of terrestrial cave faunas during such operations. An exception was made in the case of the Melones Cave Harvestman, *Banksula melones*, which was the subject of a unique rescue attempt in the 1970s when its cave habitat bordering the Stanislaus River in central California became threatened by a dam project and quarrying. In an operation involving the U.S. Army Corps of Engineers, the U.S. Fish and Wildlife Service and the World Wildlife Fund (USA), the entire collectable population of *B. melones*, together with "a portion of the overall fauna of McLean's Cave", was transplanted to an abandoned mineshaft above the high-water contour for the impoundment. The operation has been criticized on several counts, including the unsuitability of the transplant site and uncertainty as to whether caves constitute the species' principal habitat. One of the only two known cave sites in Britain of the troglobitic spider *Porrhomma rosenhaueri* is similarly threatened by quarrying, but so far no consideration has been given to the possibility of a similar rescue operation here.

Terrestrial cavernicoles may be menaced by changes in drainage patterns, sedimentation and chemical and organic pollution of cave streams, but also by other factors. Howarth has noted possible harmful effects from cigarette-smoke:

> "tobacco smoke contains a powerful insecticide which challenges, if not kills, many invertebrates in the relatively enclosed cave atmosphere".

The same researcher has documented instances of gross organic pollution of caves on Kaua'i Island in the Hawaiian chain, where the largest caves on the island became filled with fermenting molasses after the overlying field had been covered in sugar cane bagasse in an effort to build up soil on the surface. The incident occurred before the cave had been biologically investigated and resulted in the extinction of its entire fauna. In another case, a local slaughter house dumped the unwanted portions of hundreds, or perhaps thousands, of cow carcasses in a sinkhole, causing a complete change in the biota. The cave now supports a diverse collection of exotic carrion feeders, scavengers and predators, including such groups as earwigs, ants, spiders, isopods and millipedes. This invasion process has happened also on other Hawaiian Islands, such as the Big Island of Hawai'i, where arthropods recently introduced by man, especially household pests and soil forms (such as cockroaches, centipedes, bristle-tails and springtails, in addition to the groups mentioned above), have successfully colonized caves in recent years, presumably at the expense of established cavernicoles. The invasion process is aided by dumping of all kinds of organic rubbish in cave entrances.

In a biological study of a newly-discovered section of Otter Hole near Chepstow during 1977–8, I noted changes of a similar nature in the fauna present in an area which was popular as an eating place with visiting cavers. As more and more rubbish (half-empty tins and chocolate wrappers) accumulated around the picnic site, I noted a dramatic increase in numbers of the Collembolan *Folsomia candida* and the spider *Porrhomma convexum*, while three other collembolan species *Onychiurus schoetti*, *Cryptopygus garretti* and *Isotoma* cf. *agrelli* seemed to disappear completely. The cave has since been cleaned up by members of the Royal Forest of Dean Caving Club and it would be interesting to see to what extent the original fauna has recolonized.

Conservation of cave-roosting bats

The populations of bats generally, and in particular of cave-roosting bats, such as the Greater Horseshoe Bat, *Rhinolophus ferrum-equinum*, have undergone a spectacular decline this century, largely as a result of the disappearance of their insect prey through the over-use of pesticides in agriculture. We have already lost our largest cave-roosting species, the Mouse-eared Bat, *Myotis myotis*, and bat conservationists are engaged in a difficult struggle to prevent the loss of any more species.

Under the Wildlife and Countryside Act 1981 it is illegal for anyone without a licence intentionally to kill, injure or handle a wild bat of any species in Great Britain, to possess a bat, or to disturb a bat when roosting. Similarly, it is an offence to damage, destroy or obstruct access to any place that a bat uses for shelter or protection or to disturb a bat while it is occupying such a place. A strict interpretation of this law would make it an offence for cavers to even look at any bat encountered underground – for shining a light on it might be construed as wilful disturbance. Understandably, therefore, many cavers have taken the view that it is safest to deny all knowledge of bats in any caves they have visited. This is an unfortunate situation, as cavers had traditionally been a useful source of information to bat workers – often providing the first clues about the presence of bats in seldom-visited or newly-discovered caves.

In an initiative, the Fauna and Flora Preservation Society (FFPS), the former NCC and the Vincent Wildlife Trust collaborated to produce *Bats Underground a conservation code* (1988: available from FFPS). The code provides guidelines for conduct underground and is aimed at cavers. It has been well publicized in the caving press and should go some way towards alerting cavers to the problems facing bats and how to behave in bat roosts, and also towards improving the mutual understanding between cavers and bat-workers.

The code actively encourages cavers to report bat sightings (while explaining the undesirability of attempting to inspect the bats too closely) and suggests that bat survey leaders should enlist the help of cavers with their work. It also addresses the fear among cavers that reporting bat sightings will result

Fig. 7.3 Protective grid over a cave used by bats as a winter roost, Devon. (Chris Howes)

in future access being refused on grounds of disturbance. This is a very delicate area, for there is no doubt that a number of winter hibernacula would be best conserved by excluding cavers altogether. The problem is that as bat numbers decline generally, the importance of even occasionally-used hibernacula increases. At the same time, an increasing population of cavers puts ever greater demands on a finite supply of caves. The code does not recommend that access to any site be denied at all times, and it is entirely voluntary. It proposes a grading system for sites, giving guidance on visiting practices in relation to bat use and the nature of the site. The vast majority of sites would have unrestricted access, although it is suggested that winter visits (November to April) to many significant hibernacula should be avoided if possible.

In some cases, important cave or tunnel roosts that suffer disturbance may require grilles. These must be designed and built most carefully to avoid affecting the roost's climate, or interfering with flying bats. Grilles are generally placed right at the cave entrance to enclose the maximum amount of space within the roost site. Of course the principal requirement is that they should exclude people, and to this end high tensile, reinforcing steel bars of 25 mm diameter are generally used, making the construction fairly expensive. Bob Stebbings tells a gruesome tale which illustrates why such protection is still needed in our supposedly enlightened culture. A few years ago he came across the charred remains of a group of Greater Horseshoe Bats which had been drenched with lighter fuel as they hung in torpid winter sleep and then set on fire by a bunch of young hooligans "for fun".

Ultimately, the future survival of our bats depends on a change in their traditional public image as creatures of Draculean terror. To this end, the notable public relations work of Bob Stebbings and of Tony Hutson, among others, and the popular appeal of books such as Phil Richardson's *Bats* (Whittet Books, 1985), appear to have had a significant impact. The outlook for our remaining bat populations would seem to be hopeful.

Limestone quarrying

Limestone is now quarried principally as a source of aggregate used in road making and concrete, for use in the manufacture of cement and in the chemical and iron and steel industries.

In the past, some important caves have been destroyed by quarrying, such as the splendidly decorated Balch Cave on Mendip, and Fairy Hole in Weardale, Co. Durham. Although there have been few instances of significant caves lost to quarry expansion in recent years, the rate of limestone extraction has actually increased since 1983 and could accelerate further in the future if there is an upturn in the national economy leading to a major new road-building programme, or a revival of plans to generate electricity from coastal barrages such as that proposed until recently for the Severn Estuary. A further increase in the rate of extraction was expected with the advent of the Flue Gas Desulphurization programme planned for a number of major British power stations (see below).

The need to protect caves against destruction by quarrying is a compelling argument for designating important caves as SSSIs. A public enquiry over a proposal to extend Eldon Hill Quarry in the Derbyshire Peak District provided a good example of effective co-operation between the former NCC and

Fig. 7.4 Hobbs' Quarry at Penwyllt lies directly over the Ogof Ffynnon Ddu National Nature Reserve. Limestone quarrying poses the greatest threat to our caves. (Chris Howes)

caving organizations. The proposed extension would have severed the underground drainage routes linking the major cave systems of Peak Cavern and Speedwell Cavern with their catchment areas. A joint presentation by the Peak District National Park, NCC, NCA and BCRA contributed to the ultimate refusal of planning permission for the proposed extension. It should never be forgotten that our cavernous limestones, like any other geological resource, are non-renewable.

Acid rain, caves and flue gas desulphurization

The acidification of rainwaters as a result of the industrial discharge of sulphur dioxide and other acid-forming pollutants into the atmosphere, might be considered to be less of a problem in limestone ecosystems than elsewhere, as reaction with a limestone surface will effectively buffer the acids before they can do much damage.

Ironically, it is not so much atmospheric pollution itself which poses a threat to caves, but the search for a cure to the problem. Coal-fired power stations are among the worst offenders in the production of sulphur gases. Mounting pressure during the 1980s from our European neighbours who suffer the fall-out from our industrial activities persuaded the British Government to plan for the provision of emission control equipment at all our major coal-fired power stations by the year 2000.

Sulphur emissions can be 'scrubbed' from flue gases (a process known as 'Flue Gas Desulphurization', or FGD) by passing them over treated limestone. Essentially this replicates what happens when acid rain falls on a limestone terrain, but gets the dirty business done at source. There are over 200 FGD systems, of which the most commonly used are the Wellman-Lord (WL) process (a 'regenerable' system) and the limestone/gypsum (LG) process (a 'once-through' system). According to Dr M. Barrett (The Acid Question, *Mineral Planning* magazine, Sept. 1987) of Earth Resources Research Ltd., there is very little cost difference between the WL and LG systems. What separates them significantly is their implications for the destruction of limestone habitats in Britain. By recycling the limestone used to scrub the flue gases, the WL system uses up far less limestone – a non-renewable resource – and so necessitates less destruction of limestone scenery by quarrying.

In 1987, the Central Electricity Generating Board announced that they intended to retrofit an LG-process FGD system at the operational Drax B Power Station in North Yorkshire, and to build similar systems into the first batch of a series of new coal-fired power stations planned for the future. The LG process involves the conversion of hundreds of thousands of tonnes of pure limestone into gypsum for each year of operation. In meaningful terms, a power station the size of Drax B (1,875 megawatts) would consume 300,000 tonnes of limestone per year – an amount only slightly less than the current annual output of the Eldon Hill Quarry in Derbyshire. During its expected operational lifetime of 30–50 years, this one FGD plant alone would consume 9–15 million tonnes, and with 13 or so new or retrofit schemes being considered for implementation by the year 2000, this would require some 100–180 million extra tonnes of limestone to be quarried in that period, over and above the current extraction for established purposes.

The FGD programme effectively proposed to sacrifice a limited area of limestone habitat to improve the health of vast areas of other environments and as such was welcomed by conservationists. However, cavers and environmentalists were prepared to lobby government to change its mind over the system to be adopted, from the wasteful and destructive LG process, to the far less damaging WL process.

Since privatization of the electricity industry, some 80% of generating capability has passed into the hands of two private companies: National Power and Powergen. These companies may seek to move away from the present heavy reliance on home-produced 'dirty' coal by importing coal with a lower sulphur content or switching to natural gas as an alternative. For whatever reason, the FGD programme was eventually dropped, but serves as an example of the conflicts of interest that exist.

Cave SSSIs

In both Northern Ireland and the Republic, caves receive little if any statutory protection. In Britain the main government agencies concerned with the conservation of caves are the regional bodies replacing the former NCC (English Nature, Scottish Natural Heritage and the Countryside Council for Wales). They establish and manage National Nature reserves (NNRs) and are responsible for notifying Sites of Special Scientific Interest (SSSIs) under the Wildlife and Countryside Act of 1981. Such a notification becomes a charge on the land and remains in effect even when there is a change in ownership, assuring long-term protection. SSSIs may be established on biological or geological criteria. In the case of caves, the factors assessed in determining SSSI status may include cave biota, features of geological or geomorphological interest such as speleothems, clastic and mineral deposits, and palaeontological material such as cave bone deposits. Responsibility for archaeological remains rests with the Ancient Monuments Inspectorate of the Department of the Environment.

When designating an SSSI, the regional bodies are required to inform owners and occupiers of the nature of the scientific interest and of all 'potentially damaging operations' ('PDOs') which might adversely affect the site. The idea of PDOs is to control operations which fall outside the scope of the planning laws. Following SSSI designation, owners or occupiers must notify the regional body if they intend to carry out any operation included in the list of PDOs, and the regional body then has the choice of allowing the PDO to proceed, negotiating a management agreement, or, if this proves impossible, applying to the Secretary of State for a 'Nature Conservation Order' banning the proposed activity. At the present time, 12 standard PDOs are recognized for all designated cave SSSIs. They are:

1. Dumping, spreading or discharge of a material.
2. Afforestation or deforestation.
3. Changes in land drainage, including field under-drainage or moor-gripping.
4. Modification to the structure of water courses.
5. Water extraction, irrigation or other hydrological changes.
6. Extraction of minerals.
7. Construction, removal or destruction of roads, earthworks, walls and ditches, laying or removal of pipes, cables, etc.
8. Storage of materials in potholes, caves or their entrances.
9. Erection of building structures or engineering works.
10. Modification of natural or man-made features.
11. Removal of geological specimens.
12. Recreational activities likely to damage features of interest.

Fig. 7.5 Ingleborough in the Yorkshire Pennines – designated a Site of Special Scientific Interest for its limestone landforms and caves. (Chris Howes)

Notification of cave SSSIs began in 1949 and from the first, the NCC consulted with caving organizations over the selection of caves for designation. The present list of sites was drawn up by Tony Waltham, a geologist and veteran caver, in co-operation with the NCA and BCRA, and has recently been reviewed by Hardwick and Gunn (1990). Of the 48 cave SSSIs so far designated, 47 cite 'macrogeology' or 'hydrogeomorphology' as being a feature of primary interest, while in only one case is 'biology' cited. This is only to be expected as all 48 cave sites were notified within the NCC's Geological Conservation Review programme. Other features listed as being of primary interest are 'speleothems' (11 sites), 'clastic deposits' (8 sites) and 'mineral deposits' (3 sites). The single site listed as having biological interest is the Nant Glais Caves on the North Crop Limestones in South Wales. They comprise Ogof-y-Ci, which contains one of only two known British populations of the troglobitic spider *Porrhomma rosenhaueri* and Ogof Rhyd Sych, which is populated by blanched Trout.The boundaries of the cave SSSIs encompass over 30% of cave entrances and 70% of known cave passages in Britain. A number of other important caves referred to in this book, such as those of the Manifold Valley, Dovedale and Creswell Crags are also included within SSSIs which have been designated for other reasons, such as their surface karst or palaeontological interest.

Caves scheduled as SSSIs (lengths from Hardwick and Gunn, 1990):

Northern Pennines

1. Upper Dentdale Caves – including Ibbeth Peril, Hackergill: 22 caves in all, containing 3863 m of mapped passages.
2. Short Gill Cavern area: 11 caves, 781 m.
3. Leck Beck Head Catchment – from Aygill to Ireby, including all Ease Gill and Leck Fell: 79 caves, 70,505 m.
4. Kingsdale – from Marble Steps to Vesper, including West and East Kingsdale: 118 caves, 25,442 m.
5. Ingleborough – almost the entire area inside the roads, but including the Chapel Beck caves and excluding Moughton Scar: 237 caves, 53,000 m.
6. Birkwith Caves – from Calf Holes to Red Moss: 11 caves, 5558 m.
7. Brants Gill Catchment – from Hull Pot area to Hammer Pot: 33 caves, 14,089 m.
8. Pikedaw Calamine Caverns: 1021 m.
9. Sleets Gill Cave – including Dowkabottom: 2 caves, 3071 m.
10. Boreham Cave area: 7 caves, 3137 m.
11. Strans Gill: 3 caves, 427 m.
12. Birks Fell Caves – including Birks Wood System: 4 caves, 5261 m.
13. Dow Cave System: 5 caves, 4414 m.
14. Black Keld catchment – Mossdale and Langcliffe, but not Black Keld: 24 caves, 24,260 m.
15. Stump Cross Caves – including Mongo Gill: 8 caves, 6086 m.
16. Upper Nidderdale Caves – including Goyden, How Stean, Blayshaw Gill and Nidd Heads: 20 caves, 9520 m.
17. Cliff Force Cave – including Buttertubs.
18. Fairy Hole: 3479 m.
19. Knock Fell Caverns: 4000 m.
20. Hale Moss Caves – including Hazel Grove: 7 caves, 500 m.

Derbyshire

21. Castleton Caves – Perryfoot to Treak Cliff, Peak, Eldon: 32 caves, 16,911 m.
22. Bagshaw Cavern: 8 caves, 3651 m.
23. Poole's Cavern.
24. Upper Lathkilldale Caves – Lathkill Head area, Knotlow area and Water Icicle: 11 caves, 4616 m.
25. Stoney Middleton Caves – including Carlswark, Streaks: 14 caves, 4368 m.
26. Masson Hill Caves – including Jug Holes, Masson: 4 caves, 1166 m.

Mendip

27. Charterhouse Caves – Tynings Farm, GB, Rhino, Longwood, Manor Farm: 9 caves, 6747 m.
28. Cheddar Caves – Gough's area and Reservoir Hole: 39 caves, 2679 m.
29. Priddy Caves – Swildon's, Eastwater, St. Cuthbert's, Hunter's: 10 caves, 19,725 m.
30. Wookey Hole: 3658 m.
31. Lamb Leer Cavern: 585 m.
32. Thrupe Lane Swallet: 609 m.
33. St. Dunstan's Well catchment – including Stoke Lane, Shatter, Withyhill: 16 caves, 6845 m.

South Wales

34. Dan-yr-Ogof Caves – including Tunnel and Pwll Dwfn: 4 caves, 17,138 m.
35. Ogof Ffynnon Ddu area – including Pant Mawr: 4 caves, 38,318 m.

36. Little Neath River Caves -including White Lady and Town Drain: 5 caves, 9020 m.
37. Porth-yr-Ogof: 2200 m.
38. Nant Glais Caves – Ogof-y-Ci and Rhyd Sych: 2 caves, 1645 m.
39. Mynydd Llangattwg Caves – including Agen Allwedd, Craig-a-Ffynnon, Daren Cilau: 13 caves, 66,781 m.
40. Siambre Ddu: 45 m.
41. Otter Hole: 3352 m.

Other areas

42. Buckfastleigh Caves (Devon) – including Church Hill, Pridhamsleigh and Potter's Wood: 6 caves, 3000 m.
43. Napps Cave (Devon): 200 m.
44. Beachy Head Cave (East Sussex): 200 m.
45. Minera Caves (Clwyd) – including Ogof Dydd Byraf: 3 caves, 1070 m.
46. Alyn Gorge Caves (Clwyd) – including Hesp Alyn and Hen Ffynnonhau: 3 caves, 2414 m.
47. Allt nan Uamh Caves (Highland) – including Claonite: 10 caves, 3265 m.
48. Traligill Caves (Highland) – including Glenbain, Cnoc nan Uamh: 18 caves, 2057 m.

Databases

Staff at the Biological Records Centre at the Institute of Terrestrial Ecology (Monks Wood) are now producing a computer database from the biological records of the British Cave Research Association. The following will be recorded for each entry:

- Order, genus, species & subspecies
- Vice-county (no.), grid reference, status, altitude
- Recorder/collector, date of record, compiler
- Determiner, date of determination, date of compilation
- Stage, host/foodplant, habitat, depository
- Source, comments

Meanwhile, as a first stage, they have put together a site-related database for a small selection of taxa known to be associated with cave systems in Britain. This is held as a series of ASCII files on a single 5.25" IBM PC floppy disc. A 'British Cave Database' has been compiled by the Limestone Research Group based at Huddersfield University. It includes all known 'accessible cave passages formed by solutional erosion of limestone'. Two thousand seven hundred and ten caves fulfilled this criterion for entry and the following information is recorded for each:

- Name
- Location (NGR)
- Whether in SSSI and if so name
- Total passage length
- Depth
- Agricultural impacts (if any)

Taken together, these databases should provide a useful tool to those concerned with wildlife conservation in Britain, making it possible for the first time to determine the biological importance of a particular cave site, to identify the need for faunal surveys in particular caves, and to consider possible impacts on cave biotas when formulating management strategies for cave SSSIs, such as whether or not to approve a specific PDO application. In a wider context, the Cave Species Specialist Group of the International Union for the Conservation of Nature and Natural Resources (IUCN) is compiling information on threatened cave communities worldwide, with a view to eventual publication

Fig. 7.6 The famous 'Columns' in Ogof Ffynnon Ddu – the first cave to be designated a National Nature Reserve, in 1976. (Chris Howes)

of a Red Data Book on caves. It is to be hoped that the last decade of the 20th century will bring a change in the way cavernicoles are viewed by naturalists, from passive acknowledgement of their interest as biological curiosities, to active concern over the protection of this unique element of our native biota.

GLOSSARY

Aggressive water: Water containing an active chemical ingredient, such as dissolved carbon dioxide, which enables it to corrode limestone rock.
Allogenic water: Water which enters a cave after flowing overland along a permanent watercourse.
Anastomoses: Intricately branched and braided systems of narrow-diameter tubes which develop along joints and bedding planes in limestone as the initial stage in cave formation.
Anophthalmia: Eyelessness produced by genetic change in a previously sighted organis.
Aquifer: A water-bearing layer of permeable rock, sand or gravel.
Autogenic water: Water which passes directly from the soil surface into the cave without having travelled along a permanent surface watercourse.
Autotroph or producer: Green plant, bacterium or protist which can make energy-rich organic compounds (food) from simpler inorganic raw materials (*see* photo-autotroph, chemo-autotroph).
Bandits: Marauding cavernicolous predators and scavengers which plunder the resources of communities which inhabit guano beds, but which live mainly in other cave microhabitats.
Batellites: Species which depend for their livelihood on the presence of bats. They include parasites, phoretic species, scavengers, guanobia and their predators.
Bedding plane: Plane of weakness separating two strata of a sedimentary rock. Bedding planes often provide a pathway for water movement through limestone and may be opened out to form caves.
Benthic: Bottom-dwelling. Organisms which live on the bed of rivers, lakes or seas.
Biomass: The total weight of living matter, whether in an entire community, at a particular trophic level, or of a particular kind of organism in the community.
Biospeleology: The scientific study of the life within caves, as practiced by a Biospeleologist. Biospéléologie is the French equivalent.
Breakdown: A heap of rock filling all or part of a cave passage after the collapse of part of the walls or ceiling.
Bronze Age: Cultural period (c.4,400-2,750 BP) following the Neolithic and characterized by the first widespread use of metal artifacts made of bronze.
Brown Holes: Tidally-flooded caves which open on the shores of the Burren in Western Ireland.
Calcareous: Consisting of calcium carbonate, or lime.
Calcite: Mineral formed of calcium carbonate which is the major constituent of cavernous limestones and of most speleothems, such as stalactites and stalagmites.
Carboniferous: Period of Palaeozoic era (345-280MY BP) during which Britain and Ireland lay in the tropics and our major cavernous limestones were deposited.
Cave: In popular parlance, a natural subterranean void enterable by man. As a biological habitat, any perpetually dark, habitable void larger than 1 millimetre in diameter bounded by rock or similar inorganic materials.
Cavernicole: A species which lives in a cave habitat and can complete its life cycle there. *See also* troglobite.
Cave system: All the cavities and underground passages in a given area which are, or at one time were, interconnected.
Cenote: Pothole partly filled by a lake.
Chemo-autotrophs: Bacteria which derive their metabolic energy from chemical reduction or oxidation of inorganic substances, such as sulphur or iron minerals and are thus able to function independently of solar energy.
Chemosynthesis: Synthesis of complex organic molecules from simpler molecules using chemical, rather than solar energy (*see* chemo-autotrophs).
Clastic deposits: Sediments made up of eroded fragments of pre-existing rocks.

Clints: The raised slabs of a limestone pavement (*see* Grykes).
Column: A pillarlike speleothem, resulting from the union of a stalactite and a stalagmite into a single cave formation.
Consumer: *see* heterotroph.
Corrosion: Dissolution of rock by chemically aggressive water.
Crawl, crawlway: Cave passage with a ceiling so low that a caver has to crawl.
Cryophilic: Adjective applied to organisms which are adapted to cold conditions, are unable to tolerate warm conditions, and which therefore actively seek out a cool microclimate.
Deep-cave zone: The area of caves or mesocaverns where temperature remains constant within very narrow limits, the substrate remains moist and the air is permanently saturated with water vapour. Such conditions are favoured by troglobites. *See also* transition zone, stagnant-air zone.
Depigmentation: Loss of coloured pigment from the body surface due to genetic change.
Detritivores: Animals which feed on organic detritus: the dead bodies and waste products of animals, and the dead tissues of plants.
Devensian: *See* Lastglacial.
Diapause: Period of suspended animation which intervenes in the life cycle of some insects and may be essential for their normal development.
Doline: A more-or-less conical karstic depression on the land surface, produced by removal of rock by solution and underground drainage. Some dolines form when surface strata collapse into an already well-developed underground cavity.
Echolocation: Ability to detect obstacles (and sometimes food items) by emitting sound and intercepting and interpreting the reflected echoes. *See also* ultrasound.
Entomophagus: Feeding on the tissues of an insect.
Erosion: Removal of rock by physical abrasion. In caves, the abrasive is usually particles of harder minerals, such as quartz sand, wielded by fast-flowing water in the vadose zone.
Eucladioliths: Tube-shaped calcareous deposits oriented towards the source of light in a cave entrance and formed by encrustation of growing mosses, algae, etc, by tufa.
Faults: Fracture planes aligned across the rock strata, along which there has been some movement of one side relative to the other. Like joints, they may provide a structural control on the direction of formation of cave passages.
Flowstone: A compact deposit of calcite precipitated from flowing water. Precipitation occurs when carbon dioxide is lost from the water surface, reducing the amount of calcium carbonate which can be carried in solution.
Formation: *See* speleothem.
Glacial: Pertaining to cold period of Pleistocene era.
Green Holes: Submarine caves off the Burren coast of Ireland.
Gour pool: Pool of lime-rich water in a cave held back by a calcite dam. As water trickles over the dam it deposits lime, building it up even higher. Gours often form series of curved terraces down inclines. *See also* flowstone.
Groundwater: Water filling any hypogean space.
Grykes: The solutionally-opened cracks in a limestone pavement (*see* Clints).
Guanobia: Collective term for guanobious species which feed on the accumulated excrement, or guano, of bats, crickets or birds.
Heterotroph, or consumer: An organism that is unable to manufacture food from inorganic raw materials, and therefore feeds on ready-made organic molecules contained in the tissues of other organisms. *See also* autotroph, chemo-autotroph, photo-autotroph.
Hibernation: Method of surviving periods of low food availability by prolonged inactivity. Caves are used as hibernacula by various bats, which can control their temperature by choosing areas of the cave with an appropriate microclimate.
Hydrology: The scientific study of the properties, distribution and behaviour of water.
Hygrophilic: Adjective applied to terrestrial organisms which are adapted to moist conditions, are unable to tolerate dry conditions, and which therefore actively seek out a humid microclimate.

Hyperparasitic: Parasitic on a parasite.
Hypogean: That which lies beneath the surface of the earth, including the soil, rocks and any air-, or water-filled spaces within them. In the case of any body of water (salt, fresh, or frozen) open to the sky, the hypogean realm begins at the top of whatever medium lies underneath the water mass.
Igneous rock: A rock emplaced in molten form, which has subsequently cooled and solidified. *See* lava tube.
Interglacial: Warm period between Glacial advances. *See* Pleistocene.
Interstitial fauna: Animals inhabiting microcavernous spaces in-between particles of (usually unconsolidated) sediments, such as sand and fine gravels, or in porous rocks.
Joints: Stress fractures in sedimentary rocks whose alignment is often perpendicular to the strata. In limestone, they form, with bedding planes, the principal routes for water flow through the rock and so provide a blueprint for the development of cave passages.
Karren: Bare limestone whose surface has been etched and sculpted by corrosion.
Karst: Typical landform produced by sub-surface drainage and erosion of carbonate rocks, producing enclosed depressions, potholes, blind valleys and underground caverns. Named after a noted limestone area east of Trieste.
Lampenflora: Collective term for plants which grow around artificial lights in caves.
Last glacial: Most recent glacial period which reached a peak around 18,000BP and ended around 12,000 BP.
Lava tube: A type of cave tunnel formed when the surface of a basaltic lava flow crusts over, allowing the still-molten core to drain away under gravity. *See* igneous rock.
Limestone: Sedimentary rock composed primarily of calcium carbonate. It usually originates through the accumulation of limy remains of marine organisms. It is by far the most widespread and important cave-forming rock (*see also* karst).
Living fossils: Species known from fossil remains of great age, which survive to the present (*see* relict species).
Macrocaverns: The largest size class of cavernous habitats, defined as having a mean diameter of more than 200 millimetres, and corresponding to the popular notion of 'caves'. Macrocaverns are inhabited by a few specialised macrocavernicoles, but frequently have microclimate regimes which exclude the majority of mesocavernicoles. *See also* mesocaverns and microcaverns.
Maladie verte: Growths of the green alga Palmellococcus which at one time threatened to engulf prehistoric cave paintings at Lascaux in the French Dordogne.
Marine relict: An animal whose evolutionary ancestors lived in salt water, but became adapted to life in fresh water when its habitat was gradually raised above sea level, or was recharged by fresh water in some other way.
Mesocaverns: Cave habitats with a mean diameter from 1-200 millimetres, inhabited by mesocavernicoles: cave-adapted species which are often more-or-less amphibious, are tolerant of high atmospheric concentrations of carbon dioxide and are strongly hygrophilic and anemophobic. Mesocaverns form the principal habitat of troglobites. *See also* macrocaverns and microcaverns.
Mesolithic: Prehistoric cultural period (c.12000-6500 BP) between the Palaeolithic and Neolithic periods, marking the rise of modern man.
Mesozoic: Geological era (225-65MY BP), including the Triassic, Jurassic and Cretaceous periods, during which the dominant terrestrial vertebrates were dinosaurs.
Microcaverns: The smallest hypogean spaces, bounded by inorganic walls and with a diameter of less than 1 millimetre. They are inhabited by a specialized suite of tiny organisms, collectively known as the interstitial fauna, or microcavernicoles.
Microclimate: The conditions of temperature, humidity and air movement within a very restricted, defined area.
Microhabitat: A restricted, defineable space where environmental conditions differ from those in the surrounding area.
Miocene: Geological period (26-12MY BP) of the Tertiary era.

Moonmilk: Cheeselike amorphous masses consisting of carbonate minerals and micro-organisms occasionally found sprouting from the cave ceiling or walls, usually near an entrance.
Morphological: Pertaining to the physical shape or structure of an organism.
Mycophages: Animals which feed on fungi.
Nappe: Aquifer through which there is a flow of water.
Neolithic or New Stone Age: Prehistoric cultural period (c.6,500-4,000 BP) characterized by the widespread use of polished stone tools and a settled agricultural way of life. *See also* Palaeolithic, Mesolithic.
Niche: Ecological lifestyle of an organism, unique to that particular species.
Palaeolithic or Old Stone Age: Long prehistoric cultural period (c.500,000?-12,000 BP) characterized by the widespread use of chipped stone tools by members of nomadic, or semi-nomadic hunter-gatherer societies. Often divided into Lower Palaeolithic (when archaic types of Homo sapiens spread across Europe), Middle Palaeolithic (the age of *H. sapiens neanderthalis*), and Upper Palaeolithic (the age of Cro-Magnon Man, a race of *H. sapiens sapiens*). *See also* Mesolithic, Neolithic.
Percolation water: Water which trickles slowly down through the rock column along a diffuse system of poorly-developed conduits.
Periglacial: Lying just beyond the limits of an ice sheet: conditions such as are found in the tundras of Alaska and Siberia, with long, harsh winters and permafrost soils.
Permian: Geological period (280-225MY BP) of the Palaeozoic era.
Photo-autotroph: Autotroph which uses sunlight energy to make food from inorganic raw materials, a process called photosynthesis.
Phreas: System of flooded conduits beneath the water table.
Phytophages: Animals which feed on living green plants.
Pleistocene: Recent geological period (1.6MY-10KY BP) characterized by a sequence of cold Glacial, and warm Interglacial episodes.
Pliocene: Last geological period (12-2MY BP) of the Tertiary era.
Polje: Large, flat-floored, cliff-ringed karstic depression which floods periodically. Often fed by springs at one side and drained by sinks at the other side.
Pothole: Sheer-sided, open karstic pit, generally deeper than wide.
Quaternary: The current geological era (1.6MY-present) which includes the Pleistocene and Recent periods.
Relict distribution: Patchy distribution of relict species in widely scattered locations which gives evidence of former dominance of the group to which the species belongs, and/or which delimits former biogeographical domains since fragmented by the movements of crustal plates (continental drift).
Relict species: Species belonging to ancient groups whose fortunes have waned through geological time (*see* living fossils).
Saprophytes: Plants which obtain their energy heterotrophically from dead organic material.
Schiner-Racovitza System: A classification system which divides cavernicoles according to their supposed ecological dependency on caves.
Sinkhole: The point where a flow of water disappears into an underground cave.
Slutch caves: Caves formed by water flowing beneath the surface of peat deposits.
Spelaeodendron: Nickname given by cavers to the Goat Willow, *Salix caprea* which is associated with cave entrances in the Burren of Western Ireland.
Speleology: the scientific study of caves.
Speleothem: The general term for any mineral deposit or formation found in caves.
Stagnant-air zone: The area of caves or mesocaverns where air exchange with the surface is so restricted that gases from organic decomposition (especially CO_2 build up unusually high concentrations, temperature remains constant within very narrow limits, and the air is permanently saturated with water vapour. These are the conditions most favoured by troglobites. *See also* transition zone, deep- cave zone.
Stalactite: An icicle-like deposit of calcium carbonate which grows downward from the cave ceiling (*see* flowstone).

Stalagmite: A deposit of calcium carbonate which grows upward from the cave floor as the result of water dripping from above (*see* flowstone).
Straw: Hollow, thin-walled, tubular stalactite (like a drinking straw) which may grow to a length of several metres.
Sub-cutaneous zone: The uppermost layers of rock beneath the soil which in karstic terrains are often riddled with solutionally-opened cracks of mesocavernous dimensions which form the principal habitat of the majority of cavernicoles. *See* SUC.
Sub-fossil: A dead tissue, such as tooth, bone or shell, or the entire body of a dead organism preserved over a long period in its original form, rather than as a fossil.
SUC (Superficial Underground Compartment): A mesocavernous habitat of the sub-cutaneous zone of limestone and other rocks.
Sump: Section of cave passage which is completely filled with water.
Swallet: *see* sinkhole.
Taxonomic: Pertaining to the evolutionary relationship of an organism to other organisms, as expressed in the way it is classified by zoologists.
Tertiary: Geological era (65-2MY BP), including the Paleocene, Eocene, Oligocene, Miocene and Pliocene, a period marked by the emergence of mammals as the dominant terrestrial vertebrates.
Thermophilic: Adjective applied to organisms which are adapted to warm conditions, are unable to tolerate cold conditions, and which therefore actively seek out a warm micoclimate.
Threshold: Lit part of a cave entrance.
Till: A deposit of clay, sand, gravel or boulders dumped by a moving glacier.
Transition zone: The area of caves or mesocaverns whose microclimate is affected by climatic events on the surface, and where the fauna is seldom rich in troglobites. *See also* deep-cave zone, stagnant-air zone.
Triassic: Geological period (225-190MY BP), during which the dinosaurs emerged as the dominant terrestrial vertebrates.
Troglobite: A cavernicole which shows morphological features (such as loss of eyes and surface pigment) which suggest that it has undergone a long history of cave habitation.
Troglodyte: A human cave-dweller.
Troglophile: A cavernicole which is known to complete its life cycle in non-cave habitats as well as in caves.
Trogloxene: A species which occurs in caves, but does not complete its whole life cycle there.
Tufa: A generally crumbly, white, freshwater carbonate deposit precipitated in flowing springs as the result of loss of carbon dioxide from carbonate-saturated waters, either by inorganic degassing, or because it is removed by living cyanobacteria, algae, or higher plants.
Turlough: A low-lying type of karstic depression, found near the coast of Western Ireland, which fills or empties with water in response to tidal and groundwater levels.
Ultrasound: Sound of a frequency above 20 Kilohertz, beyond the range of human hearing. Such frequencies are used for echolocation by a range of animals, including microchiropteran bats, cave swiftlets and oilbirds.
Unsaturated zone: Zone of hypogean spaces lying above the water table.
Vadose flow: Water movement through the unsaturated zone.
Wall Association: A specific group of invertebrates which are frequently found on the walls and ceilings of the deep threshold and transition zones of caves in Britain and Europe.
Water table: The level below which all inter-connected spaces in rock, soil or sediments are filled with water.
Vauclusian spring: Fountaining spring fed from a phreatic conduit which at times receives more flow at its upstream end than it can carry at normal pressure, causing a head of water to build up.
Xylophages: Animals which eat wood.

SELECTED BIBLIOGRAPHY

Atkinson, T.C. 1971. The dangers of pollution of limestone aquifers, with special reference to the Mendip Hills, Somerset. *Proc.Univ. Bristol Spel. Soc.* 12: 281-290.

Balch, H.E. 1929. *Mendip: the Great Cave of Wookey Hole*. Clare, Son & Co., Wells.

Barr, T.C. 1967. Observations on the ecology of caves. *Amer. Naturalist 101*: 475-491.

Barr, T.C. 1968. Cave ecology and the evolution of troglobites. in: Dobzhansky, T. *et al.* (eds). *Evolutionary Biology 2*: 35-102. K. Holland.

Barr, T.C. and Kuehne, R.A. 1971. Ecological studies in the Mammoth Cave ecosystem of Kentucky. II. The ecosystem. *Annales de Speleologie. 26*: 47-96.

Bedford, B. 1985. *Underground Britain*, Willow Books/Collins. 175pp.

Botosaneanu, L. 1986. *Stygofauna Mundi*. Brill/Backhuys, Leiden.

Caumartin, V. 1963. Review of the microbiology of underground environments. *National Speleological Society Bulletin. 25*: 1-14.

Chapman, P. 1979. The biology of Otter Hole Cave, near Chepstow. *Trans. British Cave Res. Assoc. 6*: 159-167.

Christiansen, K.A. 1965. Behavior and form in the evolution of cave Collembola. *Evolution. 19*: 529-537.

Christiansen, K.A. and Bullion, M. 1978. An evolutionary and ecological analysis of the terrestrial arthropods of caves in the central Pyrenees. *National Speleological Society Bulletin. 40*: 103-117.

Cubbon, B.D. 1976. Cave Flora. Chapter 11 in Ford, T.D. & Cullingford, C.H.D. (eds). *The Science of Speleology*. Academic Press.

Cullingford, C.H.D. (ed.) 1953. *British Caving, an introduction to speleology*. Routledge and Kegan Paul. 468pp.

Culver, D.C. 1982. *Cave Life: Evolution and Ecology*. Harvard University Press. 189pp.

Coleman, J.C. 1965. *The Caves of Ireland*. Anvil Books.

Coleman, J.C. & Dunnington, N.J. 1944. The Pollnagollum Cave, Co. Clare. *Proc. Roy.Irish Acad.50*: 105-132.

Cox, G. & Marchant, H. 1977. Photosynthesis in the deep twilight zone: micro-organisms with extreme structural adaptations to low light. *Proc. 7th Int. Speleiol.Congr., Sheffield*: 131-3.

Davies, C. 1986. Cave- and cliff-nesting swallows in Derbyshire and the Peak. *Derbyshire Bird Report, 1986*.

Deakin, P.R. & Gill, D.W. 1975. *British Caves & Potholes*.

Deeleman-Reinhold, C.L. 1981. Remarks on the origin and distribution of troglobitic spiders. *Proceedings of the eight International Congress of Speleology, Bowling Green, Kentucky*. 1: 302-308.

Dendy, A. 1985. The cryptozoic fauna of Australasia. *Rept. Australian Assoc.Adv.Sci. 6*: 99-119.

Dickson, G.W. 1975. A preliminary study of heterotrophic microorganisms as factors in substrate selection of troglobitic invertebrates. *National Speleological Society Bulletin. 37*: 89-93.

Dickson, G.W. & Kirk, P.W. 1976. Variation in the ecology morphology and behaviour of the troglobitic amphipod crustacean *Crangonyx antennatus* Packard (Crangonychidae) from different habitats. M.S. thesis, Old Dominion University, Norfolk, Va.

Farr, M. 1980. *The darkness beckons: the history and development of cave diving*. Diadem Books.

Friederich, H., Smart, P.L. & Hobbs, R.P. 1982. The microflora of limestone percolation water and the implications for limestone springs. *Trans. Brit.Cave Res.Assoc. 9*: 15-26.

Ford, D.C., and Williams, P.W. 1989. *Karst geomorphology and hydrology.* Unwin Hyman. 601pp.

Ford T.D. 1977. *Limestones & caves of the Peak district*. Geo Books, 469pp.

Ford, T.D. (ed) 1989. *Limestones and caves of Wales*. Cambridge University Press, 257pp.

Ford, T.D. and Cullingford, C.H.D. (eds). 1976. *The Science of Speleology*. Academic Press. 593pp.

Gadeau de Kerville, H. 1926. Note sur un Protee anguillard (Proteus anguinus Laur.) ayant vecu sans aucune nourriture *Bull.Soc.Zool.France 51*.

Gledhill T. & Ladle M. Observations on the life history of the subterranean amphipod *Niphargus aquilex aquilex* Schiodte. Freshwater Biol. Assoc.

Glennie, E.A. 1947. Cave Fauna. *Cave Res.Grp. G.B. Publication 1*.

Glennie, E.A. 1967. The distribution of the hypogean Amphipoda in Britain. *Trans. Cave Res.Grp.G.B. 9*: 132-136.

Gosse, P.H. 1860. *The Romance of Natural History*. James Nisbet, London. 69-73.

Gough, J.W. 1967. *The Mines of Mendip*.

Gounot, A.M. 1967. La microflore des limons argileux souterrains: Son activite productrice dans la biocenose cavernicole. *Ann.Speleol. 22*: 23-146.

Hamilton-Smith, E. 1971. The classification of cavernicoles. *National Speleological Society Bulletin. 33*: 63-66.

Harding, P.T. and Greene, D.M. 1988. *Computerization of data on selected cave fauna in Britain. Annexes 1-8*. CSD Report no.886. Nature Conservancy Council, Peterborough.

Harding, P.T. 1989. The occurrence of Asellidae in the British Isles. Part 1. *Asellus cavaticus. Isopoda 3*: 5-7.

Hardwick, P. & Gunn, J. (Limestone Research Group, Manchester Polytechnic). 1989. *The impact of agricultural operations on the scientific interest of cave SSSI.* Unpublished report to Nature Conservancy Council, Peterborough.

Harris, J.A. 1970. Bat-guano cave environment. *Science 169*: 1342- 3.

Harrison, C.J.O. 1980. A re-examination of British Devensian and earlier Holocene Bird Bones in the British Museum (Natural History). *Journal of Archaeological Science* 7: 53-68.

Hazelton, M. & Glennie, E.A. 1953. Cave Fauna and Flora. Chapter 9 in: Cullingford, C.H.D. (ed.) *British Caving, an introduction to speleology*. Routledge and Kegan Paul.

Hazelton, M. 1955 *et seq*. **Biological records of the Cave Research Group of Great Britain.** *Biol.Suppl.Cave Res.Grp.G.B.* **Pts 1-5,** *Biol.Rec.Cave Res.Grp.C.B.* **Pts 6-8,** *Trans Cave Res.Grp.C.H. 7: 10-19, 9: 162-241, 10: 143-165, 12: 3-26, 13: 167-197, 14: 205-230, 15: 225-253*.

Hazelton, M. 1965 *et seq*. **Fauna collected from caves.** *Trans. Cave Res.Grp.G.B.7*:**26-40, 9: 242-255, 10: 167-181, 12: 75-91, 13: 198-223, 14: 231-272, 15: 203-215.**

Hazelton, M. 1974. The fauna found in some Irish caves. Irish Vice County Records of Fauna from the Hypogean and related zones of caves and Wells in Ireland. A Check List of the Irish Cave Fauna. Troglobite, Troglophile and Trogloxene, Hypogean fauna recorded from Ireland by the Cave Research Group of Great Britain 1952-1971. *Trans.Cave. Res. Grp.G.B. 15:* 191-201, 203-15, 221-2, 225-250.

Holsinger, J.R. 1981. *Stygobromus canadensis*, a troglobitic amphipod crustacean from Castleguard Cave, with remarks on the concept of cave glacial refugia. *8th Proc.Int.Congr. Speliol. Bowling Green Ky*. pp.93-95.

Howarth, F.G. 1980. The zoogeography of specialised cave animals: a bioclimatic model. *Evolution 34*: 394-406.

Howarth, F.G. 1983. Ecology of cave arthropods. *Ann.Rev. Entomol.28*: 365-389.

Howarth, F.G. 1988. Environmental ecology of North Queensland caves; or why there are so many troglobites in Australia. pp.76-84. In: Pearson, L. (ed.) *Proceedings of the 17th ASF Bienniel Conference, Lake Tinaroo, Queensland*. Australian Speleol.Federation, Cairns.

Hutson, A.M., Mickleburgh, S. & Mitchell-Jones, A.J. 1988. *Bats Underground a Conservation Code*. FFPS/NCC/Vincent Wildlife Trust.

Jeannel, R. 1943. *Les fossiles vivants des cavernes*. Editions Gallimard. 321pp.

Jefferson, G.T. 1958. A White-eyed Mutant Form of the American Cockroach, *Periplaneta americana* (L.). *Nature 182*: 892.

Jefferson, G.T. 1969. The biological work of E.A. Glennie, C.I.E., D.S.O., an appreciation. *Trans.Cave Res.Grp.G.B. 11*: 103-106.

Jefferson, G.T. 1976. Cave Faunas. Chapter 10 in Ford, T.D. & Cullingford, C.H.D. (eds). *The Science of Speleology*. Academic Press.

Jefferson, G.T. 1989. Cave Biology in South Wales. Chjapter 5 in: Ford, T.D. (ed.) 1989. *Limestones and caves of Wales*. Cambridge University Press 257pp.

Jefferson, G.T. 1983. The threshold fauna, a neglected area of British cave biology. *Studies in Speleology 4*: 53-58.

Juberthie, C. and Delay, B. 1981. Ecological and biological implications of the existence of a "superficial underground compartment". *Proc.8th Int.Congr. Speleol. Bowling Green, Ky*. pp.203-206.

Judson, D.M. and Champion, A. 1981. *Caving and Potholing*. Granada Publishing.

Judson, D.M. and Richards, L.E. 1991. Cave Conservation. Chapter 9 in: Judson, D.M. (ed.) *Caving Practice and Equipment*. David & Charles, Newton Abbot.

Keith, J.H. 1975. Seasonal changes in a population of *Pseudanophthalmus tenuis* (Coleoptera, Carabidae) in Murray Spring Cave, Indiana: a preliminary report. *Int.J.Speleol.7*: 33-44.

Kempe, D. 1988. *Living Underground*, Herbert Press.

Lankester, E.R. 1893. Blind animals in caves. *Nature 47*: 389-486.

Lavoie, K.H. 1981. Invertebrate interactions with microbes during the successional decomposition of dung. *8th Proc.Int.Congr.Speleol., Bowling Green, Ky*. pp.265-266.

Lawrence, P.N. 1960. The discovery of *Onychiurus schoetti* (Lie-Pettersen, 1896) *sensu* Stach, 1947 (Collembola) in British Caves. *Entomologist 93*: 36-39.

Lowndes, A.G. 1932. Occurrence of *Bathynella* in England. *Nature 130*: 61-62.

Magniez, G. 1978. Quelques problemes biogeographiques, ecologiques et biologiques de la vie souterraine (1). *Bull.sc.Bourg*.31: 21-35.

Maitland, P.S. 1962. *Bathynella natans*, new to Scotland. *Glasgow Naturalist 18*: 175-176.

Martel, E.-A. 1894. *Les Abimes: les eaux souterraines, les cavernes, les sources*. Lib. Charles Delagrave, Paris.

Mason-Williams, A. and Benson-Evans, K. 1967. Summary of results obtained during a preliminary investigation into the bacterial and botanical flora in caves in South Wales. *International Journal of Speleology 2*: 397-402.

Mason-Williams, A. & Holland, L. 1967. Investigations into the "wall-fungus" found in caves. *Trans.Cave Res.Grp.G.B. 9*: 137-9.

Matile, L. 1970. Les dipteres cavernicoles. *Ann. Speleo. 25*: 179- 222.

Mohr, C.E. 1976. *The world of the bat*. J.B. Lipincott.

Mohr, C.E. and Poulson, T.L. 1966. *The Life of the Cave*. McGraw-Hill, 232pp.

Moseley, C.M. 1970. The fauna of caves and mines in the Morecambe Bay area. *Trans.Cave Res.Grp.G.B. 12*: 43-56.

Oldham, T. 1975. *The Caves of Scotland*.

Peck, S.B. 1976. The effect of cave entrances on the distribution of cave inhabiting terrestrial arthropods. *International Journal of Speleology 8*: 309-3231.

Peck,, S.B. 1990. Eyeless Arthropods of the Galapagos Islands, Ecuador: Composition and Origin of the Cryptozoic Fauna of a Young, Tropical, Oceanic Archipelago. *Biotropica 22*: 366-381.

Perkins J.W. The limestone of south-west Devon. Plymouth caving group.

Piearce, T.G. & Cox, M. 1977. Distribution and response to light of unpigmented and pigmented *Gammarus pulex* L. (Crustacea, Amphipoda). *Proc.7th Int.Congr.Speleol., Sheffield* pp.351-3.

Poulson, T.L. 1964. Animals in aquatic environments: animals in caves. pp.749-771. in: Dill, D.B. (ed.) *Handbook of Physiology*. Washington D.C. American Physiol. Society.

Poulson, T.L. 1972. Bat guano ecosystems. *National Speleological Society Bulletin 34*: 55-59.

Poulson, T.L. & Kane, T.C. 1981. How food type determines community organization in caves. *8th Proc.Int.Congr.Speleol., Bowling Green, KY.* pp.56-59.

Poulson, T.L. and White, W.B. 1969. The cave environment. *Science 165*: 971-981.

Proudlove, G.S. 1979. Fishes in British caves, an interim report. *Journal Craven Pothole Club. 6*: 10-12.

Pugsley, C. 1981. Ecology of the New Zealand glowworm *Arachnocampa luminosa* (Diptera, Mycetophilidae) in caves at Waitomo, New Zealand. *8th Proc.Int.Congr. Speleol., Bowling Green, Ky.* pp.483-88.

Racovitza, E.G. 1907. Essai sur les problemes biospeleologiques. *Arch.Zool.exp.et gen. 36*: 371-488.

Ransome, R. 1980. *The Greater Horseshoe Bat.* Blandford Press.

Roe, D.A. 1981. *The Lower and Middle Palaeolithic Periods in Britain*. Routledge and Kegan Paul.

Schiner, J.R. 1854. Fauna der Adelsberger, Luegcr, und Magdalenen-Grotte, in Schmidl, A. *Die Grotten und Hohlen von Adelsberg, Lueg, Planina und Laas*. Braumuller, Vienna: 231-272.

Self, C.A. (ed.) 1981. *Caves of County Clare.* University of Bristol Spelaeological Society. 225pp.

Shaw T.R. 1974. A short history of speleology up to 1900. *Trans.Brit.Cave Res.Assoc. 1*: 1-13.

Shaw T.R. 1979. *History of cave science.* Anne Oldham, Crymych, Wales.

Sheppard, E.M. 1971. *Trichoniscoides saroeensis* Lohmander, an isopod crustacean new to the British fauna. *International Journal of Speleology 3*: 425-432.

Sket, B. 1986. Evaluation of some taxonomically, zoogeographically, or ecologically interesting finds in the hypogean waters of Yugoslavia (in the last decades). *9.Congr.Int.Espeleol.Comun. Barcelona. 2*: 125-128.

Smith, D.I. and Drew, D.P. (eds.) 1975. *Limestones and caves of the Mendip Hills.* David & Charles. 424pp.

Snazell, R. 1978. *Pseudomaro aenigmaticus* Denis, a spider new to Britain (Araneae: Linyphiidae). *Bull.Br.arachnol.Soc. 4*: 251-253.

Snazell, R. and Duffey, E. 1980. A new species of *Hahnia* (Araneae, Hahniidae) from Britain. *Bull.Br.arachnol.Soc. 5*: 50-52.

Standen, R. 1910. Notes on the cave spider, *Meta menardi* Latreille. *The Lancashire Naturalist 2*: 185-189.

Stebbings, R.E. 1988. *The Conservation of European Bats.* Christopher Helm, London.

Stuart, A.J. 1982. *Pleistocene vertebrates in the British Isles.* Longman. 212pp.

Sutcliffe, A.J. 1985. *On the track of Ice Age mammals.* British Museum (Natural History). 224pp.

Sweeting, M.M. 1973. *Karst Landforms.* Columbia Univ.Press.

Tattersall, W.M. 1930. *Asellus cavaticus* Schiodte, a blind isopod new to the British fauna. *J.Linn.Soc.Zool.37*: 79-91.

Thines, G. and Tercafs, R. *Atlas de la Vie Souterraine. Les Animaux Cavernicoles.* **de Visscher, Brussels.**

Turk, F. 1967. The non-aranean arachnid orders and the myriapods of British caves and mines. *Trans.Cave Res.Grp.G.B. 9*: 142-161.

Turk, F. 1972. Biological notes on Acari recently recorded from British caves and mines with descriptions of three new species. *Trans.Cave res.Grp.G.B. 14*: 187-194.

Tuttle, M.D. and Stevenson, D.E. 1978. Variation in the Cave Environment and its Biological Implications. In: Zuber, Chester, Gilbert and Rhodes (eds.) *1978 Proceedings of the National Cave Management Symposium.* Speleobooks, Albuquerque, New Mexico, USA.

Tratman, E.K. (ed.) 1969. *The caves of north-west Clare, Ireland*. David & Charles. 256pp.

Vandel, A. 1965. *Biospeleology, the biology of cavernicolous animals.* Pergamon Press. 524pp.

Vire, A. 1904. La Biospéléologie. *Comp. rend. Acad. Sci. Paris CXXXIX.*

Waltham, A.C.(ed.) 1974. *The limestones and caves of north-west England.* David & Charles. 477pp.

Waltham, A.C. 1983. *Caves.* Macmillan, London. 240pp.

Waltham, A.C. 1983. A review of karst conservation sites in Britain. *Studies in Speleology 4*: 85-92.

Wolf, B. 1934-1938. *Animalium Cavernarum Catalogus, pars 1-13.* Junk, Berlin.

Wymer, J. 1982. *The Palaeolithic Age.* Croom Helm.

Yalden, D.W. and Morris, P.A. 1975. *The lives of bats.* Quadrangle/New York Times Book Co.

INDEX

Page references in bold refer to illustrations